Steven J. Brams' *Theory of moves*, though based on the classical theory of games, proposes major changes in its rules to render it a truly dynamic theory. By postulating that players think ahead not just to the immediate consequences of making moves, but also to the consequences of countermoves to these moves, counter-countermoves, and so on, it extends the strategic analysis of conflicts into the more distant future. It elucidates the role that different kinds of power – moving, order, and threat – may have on conflict outcomes, and it also shows how misinformation, perhaps caused by misperceptions or deception, affects player choices. Applied to a series of cases drawn from politics, economics, sociology, fiction, and the Bible, the theory provides not only a parsimonious explanation of their outcomes but also shows why they unfolded as they did. This book, which assumes no prior knowledge of game theory or special mathematical background, will be of interest to scholars and students throughout the social sciences as well as individuals in the humanities and the natural sciences.

Steven J. Brams is professor of politics at New York University. He is the author, co-author, or co-editor of twelve previous books that involve applications of game theory or social choice theory to voting and elections, international relations, and the Bible and theology.

Theory of moves

Theory of moves

Steven J. Brams

Department of Politics
New York University

CAMBRIDGE
UNIVERSITY PRESS

Published by the Press Syndicate of the University of Cambridge
The Pitt Building, Trumpington Street, Cambridge CB2 1RP
40 West 20th Street, New York, NY 10011-4211, USA
10 Stamford Road, Oakleigh, Melbourne 3166, Australia

First published 1994

Printed in Great Britain at the University Press, Cambridge

A catalogue record for this book is available from the British Library

Library of Congress cataloguing in publication data

Brams, Steven J.
Theory of moves / Steven J. Brams.
p. cm.
Includes bibliographical references.
ISBN 0 521 45226 0 – ISBN 0 521 45867 6 (pbk.)
1. Social sciences – Mathematical models. 2. Game theory.
I. Title.
H61.25.87 1994
300'.1'5118–dc20 93-8056 CIP

ISBN 0 521 45226 0 hardback
ISBN 0 521 45867 6 paperback

CE

"We discussed what the Soviet reaction would be to any possible move by the United States, what our reaction with them would have to be to that Soviet reaction, and so on, trying to follow each of those roads to their ultimate conclusion."

Theodore C. Sorensen about the deliberations of the Executive Committee during the October 1962 Cuban missile crisis (Holsti, Brody, and North, 1964, p. 188)

"What are you going to do after that? Look ahead, look ahead – two or three or four steps ahead."

Mikhail S. Gorbachev to emissaries of the junta that staged the August 1991 attempted coup in the Soviet Union (Clines, 1991)

"Yeltsin is great in a crisis because a crisis requires a tactician. But he is not a strategist: he never thinks about tomorrow."

Pavel Voshchanov, a journalist who served as Boris N. Yeltsin's press secretary (Bohlen, 1992)

Contents

List of figures

Acknowledgments

It is a pleasure to thank my co-authors of articles, Ann E. Doherty, Marek P. Hessel, Walter Mattli, Ben D. Mor, Matthew L. Weidner, and Donald Wittman, on which part of this book is based. Ben Mor, especially, stimulated me to develop a more coherent theory of moves. I also thank D. Marc Kilgour and Frank C. Zagare for the significant help they have given me over the years in the development and application of the theory, both in their own work and in our joint research. In addition, Morton D. Davis, Alan D. Taylor, and William S. Zwicker offered valuable suggestions. John Haslam and Anne Rix at Cambridge University Press deserve special praise for their professionalism and good cheer.

The detailed comments on the entire manuscript of Zagare, Jean-Pierre P. Langlois, Julie C. Brams, and anonymous reviewers proved extremely helpful in revising it for publication. It is not unusual to thank former students, but in this instance I am especially indebted to several who wrote class papers from which I have adapted case studies: Christina Brinson, Jasmin C. Jorge, Brian J. Lahey, Bhashkar Mazumder, and John M. Parker. I am also grateful for the financial support I have received from the National Science Foundation and the C. V. Starr Center for Applied Economics at New York University on various projects that relate, directly or indirectly, to this book. Finally, the support of my family, especially my wife, Eva, has been invaluable.

Introduction

1 Overview

The theory of moves (TOM) brings a dynamic dimension to the classical theory of games, which its founders characterized as "thoroughly static" (von Neumann and Morgenstern, 1944; 3rd edn, 1953, p. 44).[1] I illustrate this dimension with two games in section 2 of this Introduction, which distinguish TOM not only from classical game theory but also from modern approaches to the study of game dynamics that use the past to explain or predict the unfolding of the future.

By postulating that players think ahead not just to the immediate consequences of making moves but also to the consequences of countermoves to these moves, counter–countermoves, and so on, TOM extends strategic thinking into the more distant future than most other dynamic theories do. By elucidating the rational flow of moves over time, it facilitates the dynamic analysis of conflicts in which thoughtful and intelligent (but not superintelligent!) players, with at least some degree of foresight, might find themselves engaged. The theory has a number of features worth noting at the outset.

Tractability

To keep the analysis tractable, I concentrate on two-person games in which each player has only two strategies and can strictly rank the resulting four outcomes from best to worst. There are exactly 78 such "strict ordinal" 2×2 games – so called because the players order the outcomes but do not attach cardinal utilities to them – but I focus on the 57 "conflict games" in which there is no mutually best outcome.

The theory is built around three basic concepts:

[1] Historical information on game theory, and biographical information on von Neumann, can be found in Luce and Raiffa (1957), Heims (1980), Poundstone (1992), Macrae (1992), and Weintraub (1992). Parts of this chapter, section 6.4, and the Appendix are adapted from Brams and Mattli (1993) with permission.

- "nonmyopic equilibria," or the stable outcomes induced when the players think ahead;
- outcomes induced when one player has "moving power," "order power," or "threat power";
- incomplete information, either about player preferences or the possessor of power in a game.

All games have at least one nonmyopic equilibrium, which the classical theory does not guarantee for Nash, or myopic, equilibria unless one admits "mixed" strategies. Such strategies, which involve randomized choices, do not lead to specific – much less unique – predictions, which some theorists contend is essential to making a theory testable and useful (Harsanyi, 1977; Harsanyi and Selten, 1988; Güth and Kalkofen, 1989).

My view is that if there are multiple equilibria, the selection of one should be sensitive to the starting point of play and who moves first (chapter 1). TOM's nonmyopic equilibria depend on these parameters, which is another way of saying that the history of play matters.

I compare nonmyopic equilibria with Nash equilibria, which are the standard equilibria in noncooperative game theory.[2] In addition, I show how the possession of moving, order, or threat power can sometimes upset Nash or nonmyopic equilibria – or stabilize nonequilibria – when there is an asymmetry in the capabilities of the two players because of a power imbalance. (Perhaps because game theory has been more developed in economics than in political science, it has been mostly oblivious to the role of power in human affairs.) Finally, I indicate how incomplete information may lead to misperception and even deception in certain situations.

Applicability

To build a foundation for the theory of moves, I postulate radical changes in the rules of play of classical game theory that, in my opinion, have strong intuitive appeal. These changes make it possible to employ TOM to explicate and interpret the rational calculations that players make in many real-life situations.

As evidence for this point, I draw on a wide variety of cases, especially

[2] *Noncooperative game theory* assumes that agreements not in the players' interest are unenforceable, whereas *cooperative game theory* assumes that they are. Provided agreements are enforceable (e.g., by the courts or some other agency), the cooperative theory is used to analyze how the rewards from forming coalitions might be reasonably apportioned among their members. TOM is part of noncooperative game theory, because it assumes that players plan their moves without presuming that other players can be bound by agreements to cooperate; rather, other players will cooperate only if they find it in their interest to do so.

in politics but in other domains, including fiction and the Bible, as well. These cases are invoked not simply to provide sketchy illustrations of the ideas of the theory but rather to depict in some detail how rational players arrive at the outcomes that they do, sometimes through blow-by-blow accounts of their calculations and actions. I also examine certain theoretical and philosophical issues (e.g., the rationality of believing in a superior being) that TOM, in my opinion, illuminates better than a myopic perspective.

Systematic results

By analyzing all possible ordinal configurations in which two players, each with two strategies, may find themselves embedded, I am able to give systematic results, based on only a few rules of play, for all 2×2 strict ordinal games. This analysis, which is deductive, uncovers a number of regularities, some of which are quite subtle. For example, I show when it is rational for a player to be magnanimous by moving, for example, from a best to a next-best outcome, lest it end up even worse if it refuses to budge from the initial outcome. I will not try to preview other findings here but instead would stress that the insights the theory offers in strategic situations as simple as 2×2 games are often compelling and, by and large, substantiated by the empirical cases.

A starting point

In game theory, a clear demarcation is made between games in "normal form," which are described by payoff matrices, and games in "extensive form," which are described by game trees. Although one form can be translated into another, game trees better capture the sequential nature of player choices, which are suppressed in a payoff matrix (its strategies describe choices contingent on other players' choices but not explicitly the order of play).

I collapse the distinction between the two different game forms here, because I am less interested in analyzing a particular sequence of choices – except in specific examples – than in exploring how the structure of payoffs in a matrix affects play of a game generally. For this purpose, I start with the payoff matrices of 2×2 games and then, because I assume that players think ahead, do game-tree analysis within the matrices.

My unorthodox mixing of the normal and extensive forms has ramifications for the play of a game. I assume that, instead of choosing strategies in a payoff matrix, whose selection by both players defines an outcome, players commence play at an outcome in a matrix, from which they then may move or not move. This point of departure gives games a beginning, endowing them with a history that helps to explain subsequent player behavior.

Specifically, players are given a basis for determining if their moves and countermoves are (nonmyopically) rational – that is, if they lead to preferred outcomes that themselves have long-term stability. By contrast, the rationality of player choices is not so apparent when players, as the classical theory assumes, select strategies *de novo* in matrix games, without considering the consequences of a series of moves and countermoves from the resulting outcome.[3]

Future horizons

Given that a series of moves and countermoves from any starting outcome is possible, it is reasonable to ask how far players think ahead in deciding whether or not to move. I assume, initially, that if they calculate that their moves will trigger a series of responses that will return them to the starting outcome, they will not move in the first place. But I abandon this assumption later and permit cycling, demonstrating how a player with moving power, by being able to force the other player to stop, can break a cycle and, on occasion, induce a better outcome for itself.[4] In the case of threat power, I assume repeated or continuing play of a game, which extends the future horizon still further.

Building blocks for larger games

Of course, even the systematic analysis of all 2×2 games does not reveal all the complexities and nuances that may occur in larger games with either more strategies, more players, or both. But some of the building blocks I provide can be extended to larger games, though not necessarily in precisely the same form that I use them here.

Indeed, I offer such an extension in chapter 7, analyzing the consequences of moves in three-party negotiations. In this model, I abandon nonmyopic calculations because the parties, who may be differently weighted, are assumed to have only incomplete information. Instead, I suppose that they successively retreat to fallback positions in order to try

[3] If players do begin by choosing strategies, I assume that they can anticipate subsequent moves in the matrix and, on this basis, play an "anticipation game." Also, in defining threat power, I assume that players choose strategies, or threaten their choice, to influence subsequent moves in the matrix game.

[4] To sidestep using either the masculine or feminine gender, or switching back and forth between them in a distracting way, I use the neuter gender for players, except when it is obviously out of place or awkward.

to improve upon the outcome. Impasse, which the parties are able to rank along with regular alternatives, occurs in the absence of agreement. Although this model differs substantially from the earlier two-person models, it illustrates how the interplay of moves can be captured in larger games, without the calculations becoming so complicated that real decision makers would throw up their hands in bewilderment.

A unified point of view

As much as it provides details on the dynamic analysis of relatively simple games, TOM offers a unified point of view on the study of strategic interaction. This view is intended as something of an antidote to the sophisticated yet often arcane game-theoretic models that adorn the literature, especially in economics, but do not paint a broad picture (Fisher, 1989). Although these models sometimes offer important insights into strategic interaction, their canvass is narrow. Worse, many are hopelessly far removed from ever being applicable to real-life situations, for either explanatory, predictive, or prescriptive purposes.

There are, of course, exceptions to such esoteric flights of fancy and, recently, efforts to make the more important findings of game theory accessible to a wider audience (e.g., Kreps, 1990; Dixit and Nalebuff, 1991; McMillan, 1992). But there is not, in my opinion, a major alternative theoretical approach that has been developed with an eye to applications and practical advice.

For the practitioner, especially, my catalogue of the properties of all the 2×2 conflict games given in the Appendix could serve the normative purpose of helping determine whether, for example, threats are effective in a particular game, and if so what kind. I believe we cannot understand the properties of larger games until we thoroughly explore the intricacies of 2×2 games.

The great virtue of these simple structures is that they enable one readily to grasp the interplay of strategic choices, which the myriad interactions of complex, detailed models tend to hide. Not only may elaborate models cloak the essential features of a problem in a fog, but they also may be "almost as opaque as the real world we wish to understand" (Richerson and Boyd, 1987, p. 33).

Fruitfulness

TOM is by no means the be-all and end-all of applied game-theoretic modeling. The dynamic analysis of ordinal games still has gaps that need to be filled and details that need to be worked out.

Because TOM is rooted only in ordinal preferences – not cardinal utilities, which are almost always impossible to ascertain in real-life situations – it is relatively easy to understand and apply. (I hope that this understanding is enhanced by my generous use of figures and descriptive aids, a Glossary of more technical terms, and the listing of games and their properties in the Appendix.) At the same time, TOM has considerable richness, allowing for different levels of anticipation, power asymmetries, incomplete information, cycling, and the like.

This richness, as well as its incompleteness and rough edges, augurs well, I believe, for TOM's fruitful further development. But I caution that the theory's mathematical extension to larger games should not be mindlessly pursued by making, for example, prodigious nonmyopic calculations on the computer. Ultimately, simplifications and generalizations will be required to make the explosion of results from such an exercise interpretable and applicable to real-life situations.

A strategy for extending the theory needs to be well thought out to give the theory breadth, depth, and applicability. I offer a glimpse of such extensions in an analysis of two larger games in section 2: a "truel," which is a three-person version of a duel; and a two-person game, wherein each player has three strategies instead of two. This brief excursion into larger games only hints at the striking features that the extant theory glosses over and TOM, when developed more systematically for larger games, may help to uncover and tie together.

2 Why a new dynamic theory?

In section 1 I listed several features of the theory of moves (TOM) that I believe make it superior to the classical theory of games, especially in modeling and prescribing solutions to real-life conflicts. But my accounting of TOM's advantages is not meant to imply that game theory in the past 50 years has made no progress in the development of a dynamic theory.

Quite the contrary. The original theory of von Neumann and Morgenstern (1944; 3rd edn, 1953) used the extensive form of a game to describe situations in which players make sequential choices in stages. Other theorists either improved upon or developed powerful theoretical tools – including backward induction, forward induction, fictitious play, Bayesian updating, Nash refinements, and theories of differential games, recursive games, repeated games, and supergames – to analyze dynamic play.

But a common problem with these approaches, at least insofar as they specify an order of play (either simultaneous or sequential choices), has been their neglect of the following question:

(1) Under what conditions would players move in the specified order?

TOM structures strategic situations so that the answer to this question emerges naturally from the theory (i.e., is endogenous), rather than being simply assumed (i.e., is exogenous). On the other hand, game theory has provided some stunningly insightful answers to the following question:

(2) When a player moves, what is it optimal for it to do?

These answers, which have been based upon different notions of optimal choice, assume that other players also make choices that affect the outcome. Thereby the choices of all players are rendered interdependent, which is the essence of a game.

Despite the impressive body of results for games in which information is incomplete, chance events occur, coalitions may form, and so on, the answers to question (2) often reflect only short-run optimality considerations. Thus, the concept of a Nash equilibrium posits an outcome to be stable when an immediate departure by any player does not lead to a better outcome, leaving another question unanswered:

(3) When players think ahead more than one step, would they act differently (say, by moving from a Nash equilibrium)?

To provide answers to questions (1) and (3), a dynamic theory, in which the order of moves is at least in part endogenous and the players are assumed to contemplate the consequences of their moves more than one step ahead, is needed.[5]

TOM begins play of a game by assuming that players are in a particular state – at which they receive payoffs – and from which they can, by switching their strategies, attempt to move to a better state. The game is dynamic because the players start with a past, defined by the state where they are presently situated, and compare this with future states they and other players can engender by moving (Basar and Olsder, 1982, p. 12).

As they look ahead at their possible moves, the possible countermoves of other players, their own counter-countermoves, and so on, the players try to anticipate where play will terminate, which I assume is when they accrue payoffs. For this purpose, TOM makes different sets of assumptions that are intended to be descriptive of different strategic situations, implications of which I develop in detail later.

In this section I give examples of two games that illustrate what I believe to be serious inadequacies in the standard theory. I have deliberately chosen games with more than two players, or more than two strategies for each player, to underscore the point that TOM is not inherently limited to 2×2 games.

[5] Other dynamic theories emphasize the process by which players move toward equilibrium by revising their strategies (e.g., by Bayesian updating), but the equilibrating process in these theories is almost invariably myopic. Later I will elaborate on this point.

A truel

A truel is like a duel, except that there are three players. Each player can either fire, or not fire, its gun at either of the other two players. I assume the goal of each player is first to survive, second to survive with as few other players as possible. Each player has one bullet and is a perfect shot, and no communication (e.g., to pick out a common target) that results in a binding agreement with other players is allowed. Here are the answers that standard game theory, on the one hand, and TOM, on the other, give to what it is optimal for the players to do in the truel:

> **Game theory:** At the start of play, each player fires at one of the other two players, killing that player.

Why will the players all fire at each other? Because their own survival does not depend an iota on what they do. Since they cannot affect what happens to themselves but can only affect how many others survive (the fewer the better, according to the postulated secondary goal), they should all blaze away at each other.[6] In the parlance of game theory, they have dominant strategies to do so.[7]

The players' strategies of all firing have two possible consequences: either one player survives (even if two players fire at the same person, the third must fire at one of them, leaving only one survivor), or no player survives (if each player fires at a different person). In either event, there is no guarantee of survival. In fact, if each player has an equal probability of firing at one of the two other players, the probability that any particular player will survive is only 0.25.

The reason is that if the three players are *A*, *B*, and *C*, *A* will be killed when either *B* fires at him or her, *C* does, or both do. The only circum-

[6] Even if the rules of the play permitted shooting oneself, the primary goal of survival would preclude committing suicide.

[7] The game, and optimal strategies in it, would change if (i) the players were allowed more options, such as to fire in the air and thereby disarm themselves, or (ii) they did not have to choose simultaneously, and a particular order of play were specified. Thus, for example, if the order of play were *A*, followed by *B* and *C* choosing simultaneously, followed by any player with a bullet remaining choosing, then *A* would fire in the air and *B* and *C* would subsequently shoot each other. (A disarmed *A* is no threat to *B* or *C*, so neither of the latter will fire at *A*; on the other hand, if *B* or *C* did not fire immediately at the other, each might not survive to get in the last shot, so they both fire.) Thus, *A* will be the sole survivor. A modified version of this scenario was recently played out in late-night television programming among the three major television networks, with ABC effectively going first with "Nightline," its well-established news program, and CBS and NBC dueling on which host to choose for their entertainment shows. Regardless of their ultimate choices, ABC "wins" when CBS and NBC are forced to divide the entertainment audience (Carter, 1992). For more on truels, see Epstein (1967, pp. 343–7), Kilgour (1973, 1975, 1978), Shubik (1982, pp. 22–5), ApSimon (1990, pp. 26–42), Dixit and Nalebuff (1991, pp. 329–31), and Silverman (1991, pp. 179–80).

stance in which *A* will survive is if *B* and *C* fire at each other, which gives *A* one chance in four. Although this calculation implies that one of *A*, *B*, or *C* will survive with probability 0.75, more meaningful for each player is the low 0.25 probability of survival.

TOM: No player will fire at any other, so all will survive.

At the start of the truel, all the players are alive, which satisfies their primary goal of survival though not their secondary goal of surviving with as few others as possible. Now assume that *A* contemplates shooting *B*, thereby reducing the number of survivors. But looking ahead, *A* knows that if it fires first and kills *B*, it will be defenseless and be immediately shot by *C*, who will then be the sole survivor.

It is in *A*'s interest, therefore, not to shoot anybody at the start, and the same logic applies to each of the other players.[8] Hence, everybody will survive, which seems a better outcome than that given by game theory's answer, in which the primary goal is not satisfied – or, quantitatively speaking, satisfied only 25 percent of the time.[9]

The purpose of TOM, however, is not to give a "better outcome" but to provide a more plausible model of a strategic situation that mimics what people might actually think and do in such a situation. In my opinion, the players of the truel, artificial as this kind of shoot-out may seem, would be motivated to think ahead, given the dire consequences of their actions.[10] Following the reasoning of TOM, therefore, they would hold their fire, knowing that if one fired first it would be the next target.

Underlying the diametrically opposed answers of game theory and TOM, which may strike one as paradoxical, is a subtle change in the rules of play that TOM implicitly introduces – namely, that players do not have to fire simultaneously at the start of play. In standard game theory, however, if the players do not fire simultaneously, an order of play is

[8] I know of no analysis of four-person shoot-outs, nor will I attempt to give them a name. But it is clear that it would be in the interest of one player to fire first, thereby setting up a duel between the other two, who will be the only players who remain armed. The situation at this point would be analogous to a truel, in which one player goes first and fires in the air (see note 7).

[9] If "primary" is taken to mean that 100 percent satisfaction of the primary goal is more important than any level of satisfaction of the secondary goal (including being the sole survivor), then TOM's answer (always surviving, but with two other players) is unequivocally better than game theory's answer (being the sole survivor, but with only a probability of 0.25).

[10] Having said this, I must note contrary evidence in the climactic scene of *Reservoir Dogs*, a 1992 film in which three professional criminals face off against each other in a spontaneous truel. But perhaps their behavior is not typical of that of most of the population, including, it is to be hoped, world leaders.

posited, in which case the player to move first – and then the later players – would not fire, given that play continues until all bullets are expended or nobody chooses to fire.

To be sure, this is the same answer as that given by TOM. The point I stress here is that the standard theory does not raise the question of which order of play the players – if, thinking ahead, they could make this choice – would adopt, given their goals.[11]

TOM, by contrast, leaves the order of play endogenous by asking of each player: Given your present situation and the situation you anticipate will ensue if you fire first, should you do so? Each player, liking its present state (all alive), which satisfies each player's primary goal, to the state that each would bring about by being the first to shoot (certain death), none shoots. Might this simple game be a metaphor of the several nuclear powers today – and why nuclear deterrence has so far succeeded?

I will not try to develop this metaphor into a model.[12] My main point is that TOM introduces an endogenous and look-ahead approach to game-theoretic analysis, which requires radical changes in the usual rules of play, as I will show later. In particular, these changes require the comparison of the past or present with the future – perhaps several steps ahead – to which the moves and countermoves may transport the players.[13]

[11] Hamilton and Slutsky (1993) do tackle this question by introducing a stage prior to actual play, at which players can choose in which of two periods to move. By "endogenizing" the order of moves in 2×2 games, they show in the resulting extended game when players will move simultaneously and when sequentially. Flexibility in the order of moves generally eliminates Pareto-inferior equilibria, except in games like Prisoners' Dilemma, echoing TOM's findings – to be presented later – on nonmyopic equilibria.

[12] For game-theoretic analyses of two-person nuclear conflict, see Brams (1985), Zagare (1987), Brams and Kilgour (1988), and Powell (1989).

[13] TOM does not assume *cheap talk*, which is costless preplay communication that does not bind the players to statements they make. In the truel, for example, a player might try to facilitate the choice of the "nobody shoots" outcome with the following statement in the preplay phase of the game: "I will not fire the first shot, but I will fire the last shot if I am alive and anybody else is." Precisely because this threat is rational to make and carry out, however, it conveys no new information that bolsters the logic of not shooting first, which is implicit in the goals of the players (if they do not have to shoot simultaneously). Except in the case of "threat power" (chapter 5), which, unlike cheap talk, presumes a binding and possibly costly commitment, I do not assume that players can communicate – including engaging in cheap talk – in the play or preplay phases of a game. Several recent game theory texts, including Rasmusen (1989), Fudenberg and Tirole (1991), and Binmore (1992), give examples of cheap talk, which can be helpful in coordination games by singling out one from among multiple equilibria.

A 3×3 game

This matrix game, shown in figure 1, is close to the games that will be analyzed in the remainder of the book, except for its larger size.[14] It has two players, Row and Column, each of which has three strategies (r_1, r_2, r_3 and c_1, c_2, c_3). The choice of a strategy by each player leads to an outcome at the intersection of these two strategies; for example, the choice of r_1 by Row and c_2 by Column leads to (4,5), giving Row a payoff of 4 and Column a payoff of 5. As with the truel, I next present the results of applying game theory and TOM to this game, finding once again that the different theories give startlingly different answers about optimal strategies:

> **Game theory:** Each player chooses a "mixed" strategy that results in "expected payoffs" of 3 for each player.

In a normal-form game like that shown in figure 1, the players are assumed to choose their strategies simultaneously or, equivalently, independently of each other. Because this game contains only three distinct sets of payoffs for the players – (0,0), (4,5), and (5,4) – each of which occurs in every row and in every column, the choice of a particular row or column would not appear to matter: whichever row or column one player chooses, each of the three outcomes can occur, depending on what strategy the other player chooses.

		Column		
		c_1	c_2	c_3
	r_1	(0,0) I	(4,5) II	(5,4) III
Row	r_2	(5,4) IV	(0,0) V	(4,5) VI
	r_3	(4,5) VII	(5,4) VIII	(0,0) IX

Key: (x,y) = (payoff to Row, payoff to Column)

Figure 1 3×3 game: payoff matrix

[14] I chose this game in part because it is a minimal example (two strategies are not enough) of a game in which no multistage learning process based on *fictitious play* – whereby the players choose strategies at each stage that are optimal against the past choices of the other player – converges to the Nash equilibrium (Shapley, 1964, pp. 24–7). Consequently, it undermines any hope of justifying Nash equilibria dynamically, whereas

Without knowledge of the other player's choice, all strategies would appear to be equally good (or bad). Furthermore, there does not appear to be a single preferred outcome, although (0,0) is definitely not preferred by the players, given that they like higher payoffs more than lower ones. However, because Row prefers (5,4), whereas Column prefers (4,5), there is a plain conflict of interest between the players over which of these two outcomes is better.

This conflict is aggravated by the fact that, given the choice of either of these outcomes – say (4,5) at r_1c_2, which I have numbered outcome II in the payoff matrix – Row would have an incentive to depart to r_3, which would give outcome VIII, or (5,4), which Row prefers. Indeed, from any (5,4) or (4,5) outcome in the payoff matrix, one player will always be motivated to switch its strategy to do better, holding the strategy of the other player constant.

A (0,0) outcome, from which both players will be motivated to depart, is even more fragile. Clearly, no pair of strategies associated with any of the nine outcomes is in equilibrium, because at least one player can always do better by departing.

The standard solution to this game is that instead of choosing any of their three *pure strategies* (with certainty), the players will choose *mixed strategies*, selecting each of their pure strategies with probability 1/3.[15] Assume, for example, that Row does this: it randomly chooses r_1 with probability 1/3, r_2 with probability 1/3, and r_3 with probability 1/3. Then it does not matter what Column does. For example, if Column chooses c_1, then on the average it will receive a payoff of 0 with probability 1/3, 4 with probability 1/3, and 5 with probability 1/3, or an *expected payoff* of 3. Likewise, if Column chooses a (1/3,1/3,1/3) mixture, the expected payoff of Row will also be 3.

The advantage that a mixed strategy has over a pure strategy is that if, say, Row chooses it, Column is prevented from doing better than 3 by selecting any other strategy (pure or mixed), which is to say that Column has no incentive to depart from any strategy. Likewise, Row has no incentive to depart from any strategy if Column uses a (1/3,1/3,1/3)

TOM provides a dynamic justification for the stabilty of two of the three distinct outcomes in this game. Recent results on the convergence, or lack thereof, of dynamic evolutionary processes are given in Samuelson and Zhang (1992).

15 Mixed strategies were originally applied only to two-person constant-sum games, according to the so-called Minimax Theorem, which I will say more about in chapter 2. In variable-sum games like the 3 × 3 game, in which the sum of the payoffs to the players at the different outcomes is not constant, there is no generally accepted solution, though the one I discuss is the one that theorists generally agree best captures the notion of a stable outcome in games. For a proof that the (1/3,1/3,1/3) mixed strategy is unique in the 3 × 3 game, see Aumann (1989, pp. 107–8); for a discussion of different interpretations of mixed strategies, see Rubinstein (1991, pp. 912–5).

mixture. The fact that the choice of this mixed strategy by both players robs each of any incentive to choose a different strategy means that the two mixed strategies are stable against each other, which defines a *Nash equilibrium*: neither player would have an incentive to depart from such a strategy pair, because it would do no better – or worse, for that matter, in this case – if it did.

That the unique Nash equilibrium in this game is mixed, and results in a worse outcome for both players' than either (4,5) or (5,4), may strike some as an unreasonable "solution." But the players' conflict over the these two outcomes vitiates the certain choice of either as stable: if one of the (4,5) or (5,4) outcomes were chosen, the player who prefers the other could do better by switching to a different strategy.

But what about the possibility of the players' coordinating their strategy choices, alternating between (4,5) and (5,4) in repeated play, or choosing one at random in one-shot game, thereby giving each player an expected payoff of 4½? The problem with this proposal is that it assumes that the players can communicate, would agree to coordinate their choices in a particular manner, and that this agreement would be binding, which is precluded in the standard (noncooperative) theory in which players are assumed to choose their strategies independently of each other, and agreements are not binding.

> **TOM:** The players will always receive payoffs of either 4 or 5, depending on where play starts and who moves first.

The point of departure of TOM is that the players are at an outcome, which is called a state, and compare their payoffs there with what they can expect to achieve by moving from it, given the possibility that the other player will move from that subsequent state, and so on. In the truel, of course, the starting point was that all three players were alive. In the 3×3 game, by contrast, there is no obvious candidate for an initial state.[16]

In the absence of such a candidate, TOM postulates that any one of the three distinct states might be the initial state. It then asks whether a player would move to another state by tracing out the consequences, in terms of payoffs, of all possible moves and countermoves from that initial state.

This was easy in the case of the truel, because the game had to end after every player had fired its bullet. What I showed was that well short of this point – in fact, before a single bullet was fired – the players would desist from firing, because if any single player fired first and killed another player, it would then be killed by the third player.

In the case of the 3×3 game, play under TOM starts at an initial state,

[16] Of course there may be such a candidate, depending on the situation being modeled. For example, if the 3×3 game models a situation in which, sadly, both players find themselves caught in one of the three (0,0) states, possible moves only from this state need be analyzed. I give several examples later in which there is an evident initial state.

with one player moving, then the other, in strict alternation. I make the simplifying assumption that if play

(i) returns to the initial state,
(ii) moves to a different state that gives the first-moving player the same payoff as the initial state and the same choices as it had before (i.e., it is *payoff equivalent*), or
(iii) moves to any other state that is payoff equivalent to a state chosen earlier,

play will terminate at this state, making it *terminal*. (The rationale of this assumption is that players will face the same consequences in the terminal state as they did in the earlier state, so they will have no reason to continue moving and only repeat themselves again.)[17] I assume that play will also terminate if either player receives its best payoff when it has the next move.

To illustrate these assumptions in the 3 × 3 game, consider the game tree in figure 2, in which I assume that the initial state is II in figure 1, giving the players payoffs of (4,5). The tree "grows" downward in figure 2, starting from the payoff of (4,5). Row, which is the only player that will have an incentive to move from (4,5) since Column receives its best payoff in this state, can choose to move from strategy r_1 either to r_2 (left branch) or r_3 (right branch), bringing it either to state (0,0) or state (5,4).

From each of these states, Column can then move next to two other states. On the left side, for example, consider the consequences of Column's moving from (0,0) either to (5,4) by choosing c_1 or (4,5) by choosing c_3. If Column moves to (5,4), Row would receive its best payoff and therefore would have no reason to move subsequently, which I call *payoff termination*. On the other hand, if Column moves to (4,5) on the right side, this would bring Row to state VI, which is payoff equivalent to state II, where play started, which I indicate by "VI = II" below this state. Since I assumed earlier that play would terminate at a state payoff equivalent to one chosen earlier (including the initial state, as here), there would be what I call *repetition termination* at state (4,5).

When Row chooses r_3 and goes down the right branch from initial state II, the story is similar, except that two of the three terminal states on this side extend one branch lower than on the left side. Reading the terminal states from left to right at the bottom of branch r_3, there is repetition termination at (4,5) (state VII is payoff equivalent to state II), there is repetition termination at (5,4) (state III is payoff equivalent to state VIII, which occurred two branches above), and there is payoff termination at (4,5) (Column will not move from its best state).

[17] The analogous but more draconian rule in chess is that a player who repeats a sequence of moves three times loses.

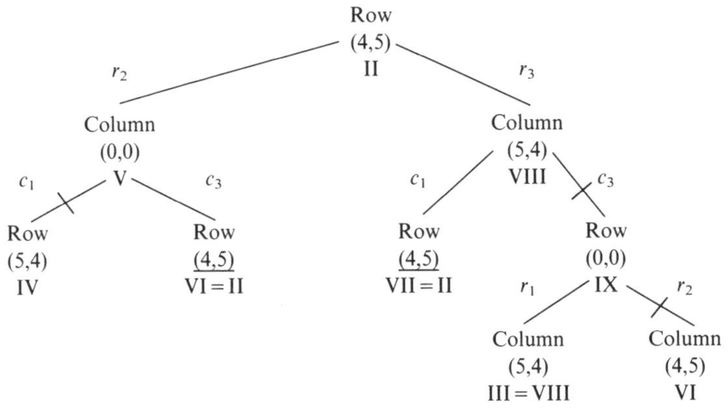

Key: (x,y) = (payoff to Row, payoff to Column)
Cut branches will not be chosen
Underscored state (VI or VII) will be chosen, following a path of uncut branches (r_2c_3 or r_3c_1), when Row moves first

Figure 2 3×3 game: Row moves from (4,5) at state II

Having found the terminal states, a procedure known as "backward induction" can now be used to analyze the game tree in figure 2. One starts at the bottom of the tree and asks, for each binary choice of a player, which strategy will lead to a better state. On the right side, for example, Row at state IX would prefer (5,4) to (4,5), so it would choose r_1 to achieve (5,4), given play reaches this state. Therefore, the branch associated with strategy r_2 is cut, indicating that Row would not choose r_2.

Knowing that (5,4) would be chosen if play reached state IX, Column at the next-higher branch must then choose between c_3 that yields this state and c_1 that gives (4,5). Preferring (4,5), Column cuts the c_3 branch that leads ultimately to (5,4).

Thus, if Row chose r_3 initially at the top of the tree, the state that would survive would be (4,5), following along the uncut path, r_3c_1, that leads to state VII, which I have underscored. Although the analysis is "backward" – from the bottom of the tree to the top – the actual choices of the players go down the tree, reflecting the sequence in which the players make them.

On the left side of the tree, starting at the bottom, Column would choose (4,5) over (5,4) in state VI, so c_1, leading to (5,4), would be cut. Hence, (4,5) will also be the survivor on the right side, demonstrating that Row's choice of either r_2 or r_3 leads to (4,5). Because Row in state II is already receiving a payoff of 4, however, I assume that Row would not move from this state only, eventually, to achieve the same outcome.

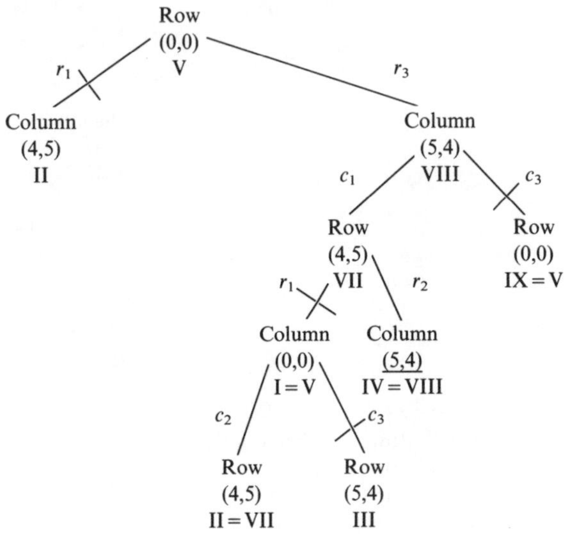

Key: (x,y) = (payoff to Row, payoff to Column)
Cut branches will not be chosen
Underscored state (IV) will be chosen, following the path of uncut branches ($r_3c_1r_2$), when Row moves first

Figure 3 3×3 game: Row moves from (0,0) at state V

Consequently, if the initial state is (4,5), Row will not move from it. Neither, of course, will Column, because it receives its best payoff in this state. Likewise, neither player will move from (5,4) if this is the initial state, making these states what I call "nonmyopic equilibria" in TOM.

But what if the initial state is, say, V, giving the players payoffs of (0,0)? The game-tree analysis in figure 3 shows that the state that would survive the backward-induction process, assuming Row moves first, is (5,4) in state IV, giving Row its highest payoff. By the same token, if Column moved first from (0,0) in state V, it could induce (4,5) in state IV. In other words, the player who moves first from (0,0) does better, establishing that each player will have an incentive to get the jump on the other when play commences at (0,0).

Later I will show that such an indeterminacy, which TOM elucidates because the order of moves is not fixed, can be resolved by "order power." The only conclusion I draw here, however, is that one or the other of the preferred states for both players, (4,5) or (5,4), will be either stable from the start or inducible if (0,0) is the initial state.

This is a dramatically different solution than that given by the mixed-strategy solution of game theory, which affords the players expected payoffs of only (3,3). Again, however, I think it not significant that TOM's solution yields a "better outcome." More significant is that players, in many strategic situations, take account of their initial state in deciding whether or not to move from it. As in the truel, not only is a player's present (or past) situation pertinent, but so also is its ability to think ahead and anticipate the consequences of a series of moves that may be set off if it moves first.

It is noteworthy, I think, that despite the fact that the 3×3 game has only three distinct outcomes and is symmetric with respect to the players' strategies, it is not a trivial exercise to establish what outcome will be chosen, starting from any initial state. In the case of the figure 2 tree, up to three levels of moves can occur before a terminal state is reached, whereas in the figure 3 tree four levels can occur. (In the 2×2 games I shall analyze later, wherein all four states are distinct, four levels are also required.)

The examples in this section illustrate that even apparently simple games may have nonobvious strategic characteristics that standard game theory has tended to hide. In my opinion, TOM helps one to sort out, explain, and thereby better understand games in which players think carefully not only about their own options but those of others as well.

Concluding remarks

I conclude on a cautionary note that the 3×3 game helps me make. Some game theorists who read an earlier version of this book said that standard game theory was perfectly adequate to do the things that TOM purports to do. They pointed out that a game like the 3×3 game in figure 1 is completely different from the games in figures 2 and 3, so it is not surprising that their solutions are different. Moreover, I do not have to invent a new equilibrium concept to distinguish the matrix and game-tree solutions – the concept of a Nash equilibrium can work in both cases.

The main reason I have developed a new equilibrium concept is that the matrix representation is, in my view, completely inadequate to model most real-life strategic situations. Players rarely, as I argue in chapter 1, choose strategies simultaneously, which is assumed in the standard theory. Nonetheless, the matrix form, stripped of its usual interpretation, deserves to be retained because it elegantly relates moves to states through what I call a "game configuration," which is the payoff matrix unencumbered by the usual rules of play (i.e., simultaneous strategy choices).

So game theorists beware: while the payoff matrix is the cornerstone of TOM, I jettison the assumption of simultaneous strategy choices. To underscore the significant differences in both the new rules I propose and the rationality calculations, power imbalances, and incomplete information I introduce, it seemed to me edifying to define new – and what I hope are intuitively appealing – concepts. Put another way, I have not abandoned the framework of game theory, which I regard as a monumental intellectual achievement, but I have attempted to make major changes in its focus.

Of course, what clarity I think TOM brings to the strategic analysis of conflicts others may see as obfuscation, or simply "old wine in new bottles." If this is the case, then the standard theory can be reclaimed. But I would warn skeptics of TOM that if what I define as "nonmyopic equilibria" are reclaimed as Nash equilibria in the appropriate extensive-form game, they will have missed my point: I do not want to analyze a plethora of game trees but rather game configurations – that is, the more parsimonious matrix form, with both the initial state and the order of play left unspecified. It is from these configurations that I derive who, if anybody, will move first – and from what states – which are predictions that can be tested in applying the theory.

To translate nonmyopic equilibria into Nash equilibria is one thing, but to refit the power concepts, and the rules of play on which they are built, into the standard theory creates a more serious translation problem. Because these concepts and rules do not have counterparts in the standard theory, incorporating them into this theory would at best be awkward, at worst require an heroic rescue operation of the standard theory.

The patchwork adjustments that would be needed to shore up the standard theory would, I'm afraid, rob it of its coherence. Perhaps it is time, therefore, to consider the need for a new and integrative theory that, as I will attempt to show, works well in explaining strategic interactions in a smorgasbord of cases.

1 Rules of play: the starting point matters

1.1 Introduction

Classical game theory, as developed by von Neumann and Morgenstern (1944; 3rd edn, 1953), distinguishes between the *extensive form* of a game and the *normal* (or *strategic*) *form*. The extensive form is represented by a game tree, in which the players make sequential choices, not necessarily knowing all the prior choices of the other players. The normal form is represented by a *payoff matrix*, in which players independently choose *strategies*, or complete plans that specify what they will do in every *contingency* – that is, for each known choice of all the other players.

TOM makes use of both forms. A payoff matrix defines the *game configuration*, which gives the basic structure of payoffs. An example of such a structure is shown in figure 1.1, which I identify as game 56.[1]

		Column (C)		
		t_1	t_2	
Row (R)	s_1	C succeeds (2,4)	R succeeds (4,2)	← Dominant strategy
	s_2	Disaster (1,1)	Compromise (3,3)	

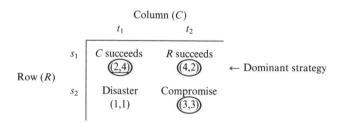

Key: (x,y) = (payoff to R, payoff to C)
4 = best; 3 = next best; 2 = next worst; 1 = worst
Nash equilibrium underscored
Nonmyopic equilibria (NMEs) circled

Figure 1.1 Game 56

Note that Row (R) has two strategies, s_1 and s_2, and Column (C) also has two strategies, t_1 and t_2, making this a 2×2 game (i.e., a game in which

[1] All different game configurations, which I shall say more about shortly, are listed in the Appendix. A roman-numeral designation will later be appended to the game numbers to indicate the initial state at which play starts.

there are two players, each with two strategies). These strategies may be thought of as alternative courses of action that the players might choose, such as to cooperate or not to cooperate.[2]

The choice of a strategy by R and a strategy by C leads to an *outcome*, with an associated payoff, at the intersection of these strategies in the payoff matrix. I assume that the players can *strictly* rank the outcomes as follows (i.e., there are no ties): 4 = best, 3 = next best, 2 = next worst, and 1 = worst. Thus, the higher the number, the greater the payoff; but these payoffs are only *ordinal*: they indicate only an ordering of outcomes from best to worst, not the degree to which a player prefers one outcome over another. That is, the payoffs describe only the *preferences* of players.

To illustrate, if a player despises the outcome it ranks 1 but sees little difference among the outcomes it ranks 4, 3, and 2, the "payoff distance" between 4 and 2 will be less than that between 2 and 1, even though the numerical difference between 4 and 2 is greater.[3] Games in which players strictly rank outcomes from best to worst are called *strict ordinal* games.

Assume R chooses s_1 and C chooses t_1 in the figure 1.1 payoff matrix. The resulting outcome is that shown in the upper left-hand corner of the matrix, with payoffs of (2,4) to the players. By convention, R's payoff is the first number in the ordered pair (2) and C's is the second (4), so R receives its next-worst payoff and C its best payoff. As shorthand verbal descriptions, I call (2,4) "C succeeds," (4,2) "R succeeds," (3,3) "Compromise," and (1,1) "Disaster."

In section 1.2, I will analyze game 56 using standard game theory. After specifying TOM's rules of play relating to possible moves the players can make in section 1.3, I will then analyze game 56 according to TOM in section 1.4, where I introduce rationality rules and illustrate the use of "backward induction" by the players in order to look ahead and determine "nonmyopic equilibria."

In section 1.5 I use game 56 as a model for interpreting the conflict between Samson and Delilah in the Bible, showing that TOM offers a more natural interpretation of the outcome of this story than does the standard theory. In section 1.6 I distinguish between "feasible" and

2 In more complex games, strategies are plans representing contingent choices. For example, assume C chooses its strategy first. Then a strategy for R might be to choose s_1 if C chose t_1, and to choose s_2 if C chooses t_2. As I shall show later, the analysis of such contingent choices is incorporated in TOM via backward induction on game trees.

3 How much a player values an outcome is normally measured in "utilities," but these are not relevant to TOM's predictions, which hold for any utilities consistent with player rankings of outcomes in a game. In the figure 1.1 game, for example, assume that the utilities of the players are the same as their ranks: 4, 3, 2, and 1. Now if the two players change their valuations of their worst outcomes from 1 to, say, −100 – as the use of "despise" in the text suggests – the predictions of TOM will be the same.

"infeasible" moves and draw some conclusions about the unique perspective that TOM offers.

Remarkably, the standard theory shows game 56 to have only one equilibrium solution, (2,4), whereas TOM demonstrates that (4,2) and (3,3) may also be solutions, depending on where play of the game starts. In fact, (3,3) is not only the outcome that TOM predicts in the Samson and Delilah story – based on the starting point – but it is also the outcome that actually occurred, illustrating the need to ground the rationality of moves on the history of play.

1.2 Applying the standard theory

The standard theory I illustrate in this section is that of the normal form, in which players are assumed to make simultaneous strategy choices in a 2×2 game. (If their choices are not literally simultaneous, the normal form assumes them to be independent of each other, so neither *R* nor *C* knows the other's choices when it makes its own.) In section 1.3 I will introduce the extensive form and show how it can be applied to the analysis of game 56, based on new rules of play.

First consider what strategy it is rational for *R* to choose in game 56. If *C* selects t_1, *R* has a choice between (2,4) and (1,1) in the first column; its payoff will be 2 if it chooses s_1 and 1 if it chooses s_2. On the other hand, if *C* chooses t_2, *R* has a choice between (4,2) and (3,3) in the second column; its payoff will be 4 if it chooses s_1 and 3 if it chooses s_2.

Clearly, *R* is better off choosing s_1 whatever contingency arises – that is, whichever strategy *C* chooses (t_1 or t_2). When one strategy of a player is unconditionally better than another strategy because its superiority does not depend on the contingency, this strategy is said to be *dominant*. *R*'s strategy of s_1 is dominant, whereas its strategy of s_2 is *dominated*, or unconditionally worse than s_1, because it always leads to inferior payoffs.

By contrast, *C* does not have a dominant strategy in game 56. Its better strategy depends on *R*'s strategy choice: if *R* chooses s_1, *C* is better off choosing t_1 because it prefers (2,4) to (4,2) in the first row; but if *R* chooses s_2, *C* is better off choosing t_2 because it prefers (3,3) to (1,1) in the second row. The fact that *C* does not have an unconditionally better strategy, independent of the contingency (i.e., *R*'s choice), makes its two strategies *undominated.*

In a game of *complete information*, in which both players have full knowledge of each other's payoffs as well as their own, *C* will know that *R*'s dominant strategy is s_1. Because s_1 is always better than s_2 for R, *C* can surmise that *R* will choose s_1. Given that *R* chooses s_1, it is rational for *C* to choose t_1, yielding (2,4) as the rational outcome of the game.

Curiously, this outcome is only *R*'s next-worst (2), though *R* is the player with the dominant strategy. *C*, the player without a dominant strategy, obtains its best outcome (4).

Nevertheless, (2,4) has a strong claim to be called the solution of game 56. Not only is it the product of one player's (*R*'s) dominant strategy and the other player's (*C*'s) best response to this dominant choice, but it is also the unique "Nash equilibrium."

A *Nash equilibrium* (Nash, 1951) is an outcome from which neither player would unilaterally depart because it would do worse, or at least not better, if it did.[4] Thus, if *R* chooses s_1 and *C* chooses t_1, giving (2,4), *R* will not switch to s_2 because it would do worse at (1,1); and *C* will not switch to t_2 because it would do worse at (4,2). Hence, (2,4) is stable in the sense that, once chosen, neither player would have an incentive to switch to a different strategy, given the other player does not switch.

This is not true in the case of the other three outcomes in game 56. From (4,2), *C* can do better by departing to (2,4); from (3,3), *R* can do better by departing to (4,2); and from (1,1), either player can do better by departing, *R* to (2,4) and *C* to (3,3).

In the latter case, if both players switched their strategies – perhaps unbeknownst to each other – in an effort to scramble away from the mutually worst outcome of (1,1), they would end up at (4,2), which also is better for both. Indeed, because (4,2) is *R*'s best outcome, *R* would be the player that would most welcome a double departure; next most welcome would be a departure by *C* alone to (3,3); and least welcome a departure by *R* alone to (2,4).

C would not particularly welcome a double departure, obtaining only its next-worst payoff of 2. Like *R*, it would prefer that its adversary make the first move from (1,1), because *R*'s departure would yield (2,4) whereas *C*'s departure yields (3,3). I shall give examples later in which the opposite is true: each player would prefer to be the first to depart from an outcome rather than wait for its adversary to make the first move.

The standard theory, by assuming that players choose strategies simultaneously,[5] does not raise questions about the rationality of moving or

[4] Technically, this equilibrium is defined by the strategies (s_1 and t_1) that yield this outcome – not the outcome itself – which are "pure" in the sense that they are chosen with certainty. Strategies may also be "mixed," which means that a player chooses a strategy at random according to some probability distribution. Although Nash equilibria may be in mixed strategies, such equilibria are not defined in games with ordinal payoffs. Moreover, even if the ordinal payoffs of game 56 were assumed to be cardinal utilities, this game would not possess a mixed-strategy Nash equilibrium. However, there are 2×2 strict ordinal games whose cardinal equivalents do possess mixed-strategy equilibria, which will be discussed in sections 2.3 and 2.5 and compared with the predictions of TOM.

[5] Strategies may allow for sequential choices, but the standard theory does not make endogenous who moves first, as TOM does, but rather specifies a fixed order of play.

departing from outcomes – at least beyond an immediate departure, à la Nash. In fact, however, most real-life games do not start with simultaneous strategy choices but commence at outcomes. The question then becomes whether a player, by departing from an outcome, can do better not just in an immediate or myopic sense but, instead, in an extended or nonmyopic sense.

There are 78 2×2 strict ordinal games that are structurally distinct in the sense that no interchange of the players, their strategies, or any combination of these can transform one of these games into any other.[6] These games represent all the different configurations of ordinal payoffs in which two players, each with two strategies, may find themselves embedded.

Game 56 is only one such configuration. The rules of play I shall propose next apply to all 78 games – and, more generally, to all finite two-person games – but in this chapter I shall illustrate them only for game 56. Results for this game provide a preview of some but by no means all the results for the set of 2×2 strict ordinal games.

1.3 Rules of play of TOM

A *game* "is the totality of rules of play which describe it" (von Neumann and Morgenstern, 1953, p. 49).[7] The first four *rules of play* of TOM for

Indeed, there is a notion of equilibrium, due to Stackelberg (1934), that supposes one player (the leader) acts before the other (the follower) does. More precisely, a *Stackelberg equilibrium* is an outcome at which one player (the leader), anticipating the best response of the other player (follower) to its (the leader's) choice of a strategy, cannot obtain a better outcome for itself by choosing a different strategy. In game 56, (3,3) is the Stackelberg equilibrium if *R* is the leader: anticipating that *C* will choose t_1 if it chooses s_1, giving (2,4), and *C* will choose t_2 if it chooses s_2, giving (3,3), *R* will choose the latter strategy, making (3,3) the Stackelberg equilibrium – or, more correctly, the strategies s_2 and t_2 associated with this state. If *C* is the leader, (2,4) is the Stackelberg equilibrium, duplicating the Nash equilibrium in game 56. In repeated games and supergames (infinitely repeated games), it is postulated that repetition occurs without any change, with play starting anew at the conclusion of each round of play. Although the positions of the players may be different as a result of play in previous rounds, the strategy choices and payoffs available to them on each round are unvarying, which I believe does not capture the fluidity of many real-life situations. If history repeats itself, it does so with some interesting variations, which TOM captures in the different states to which players can choose to move. (In chapter 4, I allow players to return to earlier states, and, in chapter 5, I allow for repeated play, but these are aspects that one may or may not want to build into a model – TOM is flexible in permitting choice on this matter.)

[6] For complete listings of the 78 games, see Rapoport and Guyer (1966) and Brams (1977); for a partial listing that excludes the 21 games with a mutually best (4,4) state, see Brams (1983, pp. 173–7). A listing of the latter games is given in the Appendix.

[7] Equating a game with its rules leaves out a lot about how the play of a game gets translated into an outcome. I shall have much more to say about this question throughout this book when I discuss the implications of different sets of rules for the choice of outcomes.

two-person games, which describe the possible choices of the players at each stage of play, are as follows:

1 Play starts at an outcome, called the *initial state*, which is at the intersection of the row and column of a 2×2 payoff matrix.
2 Either player can unilaterally switch its strategy, and thereby change the initial state into a new state, in the same row or column as the initial state.[8] The player who switches is called player 1 (*P*1).[9]
3 Player 2 (*P*2) can respond by unilaterally switching its strategy, hereby moving the game to a new state.
4 The alternating responses continue until the player (*P*1 or *P*2) whose turn it is to move next chooses not to switch its strategy. When this happens, the game terminates in a *final state*, which is the *outcome* of the game.

Note that the sequence of moves and countermoves is *strictly alternating* (the possibility of backtracking will be considered in sections 5.1 and 7.1): first, say, *R* moves, then *C* moves, and so on, until one player stops, at which point the state reached is final and, therefore, the outcome of the game.

The use of the word "state" is meant to convey the temporary nature of an outcome, before players decide to stop switching strategies.[10] I assume that no payoffs accrue to players from being in a state unless it is the final state and, therefore, becomes the outcome (which could be the initial state if the players choose not to move from it).

To assume otherwise would require that payoffs be cardinal rather than ordinal, with players accumulating them as they pass through states. I eschew this assumption in part because I think payoffs to players in most real-life games cannot be quantified and summed across the states visited.

Helpful explorations of the relationship between games and their rules can be found in Hirshleifer (1985) and Gardner and Ostrom (1991).

8 I do not use "strategy" in the usual sense to mean a complete plan of responses by the players to all possible contingencies allowed by rules 2–4, because this would make the normal form unduly complicated to analyze. Rather, *strategies* refer to the choices made by players that define a state, and *moves* and *countermoves* to their subsequent strategy switches from an initial state to a final state in an extensive-form game, as allowed by rules 2–4. For another approach to combining the normal and extensive forms, see Mailath, Samuelson, and Swinkels (1993).

9 I leave open here the question of which player, *C* or *R*, *P*1 is. Instead, I assume that the order of moves is endogenous, based on a rationality calculation of the players – to be developed shortly – and not something that is dictated by a rule. (However, players may possess special prerogatives, like greater power, that helps settle the matter.) By contrast, most solution concepts in game theory do not take into account the order of moves or, for that matter, other procedural aspects of a game (Moldovanu and Winter, *n.d.*).

10 The notion of a "temporary equilibrium" in economics (Grandmont, 1988) conveys a similar idea about the ephemeral nature of states. But in this literature, the emphasis is on predicting a sequence of short-run equilibria in successive periods that lead to a long-run equilibrium, whereas a state in TOM is not in equilibrium until it is the final state.

More significant, payoffs in the games that most interest me depend overwhelmingly on the final state reached, not on how it was reached. In politics, for example, the payoff for most politicians is not in campaigning, which is arduous and costly, but in winning.

Rule 1 differs radically from the corresponding rule of play in the standard theory, in which players simultaneously choose strategies, in a matrix game, that determine an outcome. Instead of starting with strategy choices, I assume that players are already in some state at the start of play and receive payoffs from this state if they stay. Based on these payoffs, they must decide, individually, whether or not to change this state in order to try to do better.[11]

To be sure, some decisions are made collectively by players, in which case it would be reasonable to say that they choose strategies from scratch, either simultaneously or by coordinating their choices. But if, say, two countries are coordinating their choices, as when they agree to sign a treaty, the important question is what individualistic calculations led them to this point.[12] The formality of jointly signing the treaty is the culmination of their negotiations, which covers up the move–countermove process that preceded it. This is precisely what TOM is designed to uncover.

To continue this example, the parties who sign the treaty were in some prior state, from which both desired to move – or, perhaps, only one desired to move and the other could not prevent this move without hurting itself. Eventually they may arrive at a new state (e.g., after treaty negotiations) in which it is rational for both countries to sign the treaty that has been negotiated.

Put another way, almost all outcomes of games that we observe have a history; the players do not start from a *tabula rasa*. My interest is in

[11] Alternatively, players may be thought of as choosing strategies initially, after which they perform a thought experiment of where moves will carry them once a state is selected. The concept of an "anticipation game," developed in chapter 2, advances this idea, which might be considered dynamic thinking about the static play of a matrix game. Generally, however, I assume that "moves" describe actions, not just thoughts, though I readily admit the possibility of the thought interpretation.

[12] By focusing on the calculations of individual players, I shun the "cooperative" viewpoint in game theory, which, as I noted in the Introduction, assumes that players can make an agreement that is binding and enforceable. If this is the case, they need only be concerned with how to divide up the surplus that accrues from their cooperation in some equitable or otherwise reasonable manner. But their decision to cooperate in the first place, in my view, should emerge as the result of "noncooperative" individualistic calculations, which would inform them, for example, that such an agreement is stable instead of their just assuming this to be the case. Building cooperative game theory on noncooperative foundations is what is known as the "Nash program" in game theory. It is a program that I endorse and consider to be consistent with the purposes of TOM, which simply offers a different basis for making the individualistic calculations of noncooperative game theory.

explaining strategically the progression of (temporary) states that lead to a (more permanent) outcome.[13]

Of course, what is "temporary" and what is "more permanent" depends on one's time frame. I use the phrase "more permanent," rather than simply "permanent," to underscore the obvious point that nothing in the world is permanent. Less obvious, a state that persists for a week, say, in a crisis may be permanent enough to represent an outcome in the analysis of crisis behavior, whereas a week for most historians is not long enough for a situation to qualify even as a state (unless it is exceedingly eventful and gives payoffs to the players for being there).

However it is defined empirically, I start play of a game in a state, at which players accrue payoffs only if they remain in that state so that it becomes the outcome of the game. If they do not remain, they still know what payoffs they would have accrued had they stayed; hence, they can make a rational calculation of the advantages of staying or moving. They move precisely because they calculate that they can do better by switching states, anticipating a better outcome if and when the move–countermove process finally comes to rest.

Their mental calculations of advantage and disadvantage, I assume, precede and therefore serve as the basis of their actual physical moves. That is, players work out in their minds, as best they can, the consequences of their moves and then act accordingly. But, as I indicated earlier, there is no sharp distinction between mental and physical moves. Thus, for example, leaders may float a "trial balloon," which may signal an action to be taken that may be retracted if the response to the signal is negative. Although such a move, if retracted, violates the no-backtracking assumption, I discuss later how the response to such a move might define a new game, in which backtracking in the old game is no longer an issue because circumstances (in light of the new information that the trial balloon surfaced) create a new game.

To summarize, I assume that most games have a history, which starts at some initial state. The game is different when play starts elsewhere, and so are the calculations of its players, who occupy different positions (Güth, 1991; Mertens, 1991). The choice of the initial state, and what constitute future states and eventually an outcome, depends on what the analyst seeks to explain. The time perspective of most political scientists probably ranges between about a week (e.g., in analyzing a crisis) and a generation;

13 Greenberg (1990) proposes a similar but more general approach in his "theory of social situations," wherein a situation is defined by both a "position" (state) and an "inducement correspondence," which specifies for every set of players the positions it can induce from any given position. TOM, via its rules of play and rationality rules (one set of the latter rules is given in section 1.4), offers specific mechanisms by which outcomes can be induced from any state.

journalists are more likely to think in terms of hours and days; and most historians study spans varying from a few years to a century or two.

1.4 Rationality rules and backward induction

Rules 1–4 say nothing about what causes a game to end, but only when: termination occurs when a "player whose turn it is to move next chooses not to switch its strategy" (rule 4). But when is it rational not to continue moving, or not to move from the initial state? To answer this question, I posit a rule of *rational termination* (Brams, 1983, pp. 106–7), which has been called "inertia" by Kilgour and Zagare (1987, p. 94). It prohibits a player from moving from an initial state unless doing so leads to a better (not just the same) final state:

5 A player will not move from an initial state if this move
  (i) leads to a less preferred final state (i.e., outcome); or
  (ii) returns play to the initial state (i.e., makes the initial state the outcome).

I shall discuss shortly how rational players, starting from some initial state, determine what the outcome will be by using backward induction.

Condition (i) of rule 5, which precludes moves that result in an inferior state, needs no defense. But condition (ii), which precludes moves to the same state because of cycling back to the initial state, is worth some elaboration. It says that if it is rational, after *P*1 moves, for play of the game to cycle back to the initial state, *P*1 will not move in the first place. After all, what is the point of initiating the move–countermove process if play simply returns to "square one," given that the players receive no payoffs along the way (i.e., before an outcome is reached)?[14]

Not only is there no gain from cycling, but in fact there may be a loss because of so-called transaction costs that players suffer by virtue of making moves that, ultimately, do not change their situations. Therefore, it seems sensible to assume that *P*1 will not trigger a move–countermove process if it only returns the players to the initial state, making it the outcome.[15]

It should be emphasized that part (ii) of rule 5 that precludes returning

[14] In the terminology used in the Introduction, because the initial state is payoff equivalent to itself after one cycle, play will terminate after this cycle (repetition termination). Recall that repetition termination occurred in the 3×3 game not just because of cycling back to the initial state but also because the players could revisit other states as well. But in a 2×2 game, the first revisit possible is back to the initial state, which part (ii) of rule 5 prohibits. On the other hand, rule 5′ (chapter 4) sanctions cycling.

[15] The existence of transaction costs is not critical to my no-cycling argument, because even without them the players would not improve their positions by cycling. I mention such costs, which seem ubiquitous, to lend additional plausibility to part (ii) of rule 5.

to the initial state depends on *P*1's calculation about what it is in each player's interest to do. In particular, after an initial move by *P*1, a countermove by *P*2, and a counter-countermove by *P*1, it is *P*2 – not *P*1 – that is in the position to complete the cycle. Given that it is rational for the players to make the subsequent moves that lead back to the initial state after *P*1 moves initially, then part (ii) of rule 5 says that *P*1 will not move from the initial state. This calculation on *P*1's part is purely individualistic; it is not the product of a joint decision that the players reach cooperatively about not repeating themselves, which would not properly be part of noncooperative game theory.

At this point, I make rule 5 only provisional; an alternative rule (5′) that allows for cycling will be considered in section 4.2 (along with "moving power" as a way to break cycles). I call rule 5 a *rationality rule*, because it provides the basis for players to determine whether they can do better by moving from a state or remaining in it. Still other rules of play and rationality rules, which allow for threats in repeated games (chapter 5) and incomplete information in negotiations (chapter 7), will be analyzed, though they do not involve moves and countermoves within a payoff matrix.

A final rule of TOM is needed to ensure that both players take into account each other's calculations before deciding to move from the initial state. I call this rule the *two-sidedness rule*, which is another kind of rationality rule:

6 Given that players have complete information about each other's preferences and act according to the rules of TOM, each takes into account the consequences of the other player's rational choices, as as well as its own, in deciding whether to move from the initial state or later, based on backward induction. If it is rational for one player to move and the other player not to move from the initial state, then the player who moves takes *precedence*: its move overrides the player who stays, so the outcome will be induced by the player who moves.[16]

Later I will relax the assumption of complete information to take into account the possibility of misperception and even deception, which incomplete information permits.

Because players have complete information, they can look ahead and anticipate the consequences of their moves. To show how they do so and illustrate the meaning of *backward induction* in the context of 2×2 games, consider again game 56. I show below the progression of moves, starting

[16] The indeterminacy created when both players want to move, which requires "order power" to resolve, will be discussed in chapter 5.

from each of the four possible initial states and cycling back to this state, and indicate where rational players will terminate play:[17]

1 **Initial state** (2,4). If *R* moves first, the counterclockwise progression from (2,4) back to (2,4) – with the player (*R* or *C*) who makes the next move shown below each state in the alternating sequence – is as follows (see figure 1.1):[18]

	State 1		State 2		State 3		State 4		State 1
	R		*C*		*R*		*C*		
R starts:	(2,4)	→	(1,1)	→	(3,3)	→|	(4,2)	→	(2,4)
Survivor:	(3,3)		(3,3)		(3,3)		(2,4)		

The *survivor* is determined by working backward, after a putative cycle has been completed. Assume the players' alternating moves have taken them counterclockwise in game 56 from (2,4) to (1,1) to (3,3) to (4,2), at which point *C* must decide whether to stop at (4,2) or complete the cycle and return to (2,4). Clearly, *C* prefers (2,4) to (4,2), so (2,4) is listed as the survivor below (4,2): because *C* would move the process back to (2,4) should it reach (4,2), the players know that if the move–countermove process reaches this state, the outcome will be (2,4).

Knowing this, would *R* at the prior state, (3,3), move to (4,2)? Because *R* prefers (3,3) to the survivor at (4,2) – namely, (2,4) – the answer is no. Hence, (3,3) becomes the survivor when *R* must choose between stopping at (3,3) and moving to (4,2) – which, as I have just shown, would become (2,4) once (4,2) is reached.

[17] Where, of course, depends on the endstate, or *anchor*, from which the backward induction proceeds, which I assume here – for reasons given in the text – is after one complete cycle. This assumption defines a finite extensive-form game, to which many of the so-called refinements of Nash equilibria, including subgame perfection, are applicable. However, the assumption of an anchor is dropped in section 4.2, where alternative rationality rules are applied to "cyclic games," which are finite but not of specific length, because they may cycle until the player without "moving power" quits. Because when this occurs is not specified, only where (i.e., in which state), a finite extensive-form game is not well defined. Consequently, many of the Nash refinements, which are discussed in van Damme (1991b) and several recent game theory texts (Rasmusen, 1989; Myerson, 1991; Fudenberg and Tirole, 1991; Binmore, 1992), are not in general applicable to TOM. By contrast, Skyrms (1990) analyzes various refinements using a dynamic model that assumes players maximize expected utility and do Bayesian updating. Whereas TOM is nonmyopic and nonquantitative, postulating only ordinal payoffs and discrete moves between states, Skyrms' model is myopic and quantitative, postulating cardinal payoffs and continuous moves. Virtually all dynamic approaches that I know of, except those involving repeated games and supergames that I criticized on different grounds earlier, assume myopic look-ahead calculations. While I take a myopic tack in the negotiation model in chapter 7 to keep the calculations manageable (and realistic) in larger games, this model – like all others in this book – remains an ordinal one.

[18] The labeling of states 1 to 4, which I give only in this case, is included only to show how the progression of moves returns *R* to state 1 after four moves.

At the prior state, (1,1), *C* would prefer moving to (3,3) than stopping at (1,1), so (3,3) again is the survivor if the process reaches (1,1). Similarly, at the initial state, (2,4), because *R* prefers the previous survivor, (3,3), to (2,4), (3,3) is the survivor at this state as well.

The fact that (3,3) is the survivor at initial state (2,4) means that it is rational for *R* initially to move to (1,1), and *C* subsequently to move to (3,3), where the process will stop, making (3,3) the rational choice if *R* has the opportunity to move first from initial state (2,4). That is, after working backward from *C*'s choice of completing the cycle or not at (4,2), the players can reverse the process and, looking forward, determine that it is rational for *R* to move from (2,4) to (1,1), and *C* to move from (1,1) to (3,3), at which point *R* will stop the move–countermove process at (3,3).

Notice that *R* does better at (3,3) than at (2,4), where it could have terminated play at the outset, and *C* does better at (3,3) than at (1,1), where it could have terminated play, given that *R* is the first to move. I indicate that (3,3) is the consequence of backward induction by underscoring this state in the progression; it is the state at which *stoppage* of the process occurs – *R* will not move on from (3,3). In addition, I indicate *blockage* by the vertical line blocking the arrow emanating from (3,3); a player will always stop at a blocked state, wherever it is in the progression. Stoppage occurs when blockage halts play for the *first* time from some initial state, as I illustrate next.

If *C* can move first from (2,4), backward induction shows that (2,4) is the last survivor, so (2,4) is underscored when *C* starts. Consequently, *C* will not move from the initial state, where there is blockage (and stoppage), which is hardly surprising since *C* receives its best payoff in this state:[19]

	C		*R*		*C*		*R*		
C starts:	<u>(2,4)</u>	→|	(4,2)	→|	(3,3)	→	(1,1)	→	(2,4)
Survivor:	(2,4)		(4,2)		(2,4)		(2,4)		

As when *R* has the first move, (2,4) is the first survivor, working backward from the end of the progression, and is also preferred by *C* at (3,3). But then, because *R* at (4,2) prefers this state to (2,4), (2,4) is temporarily displaced as the survivor. It returns as the last survivor, however, because *C* at (2,4) prefers it to (4,2).

Thus, the first blockage and, therefore, stoppage occurs at (2,4), but blockage occurs subsequently at (4,2) if, for any reason, stoppage does

[19] But in chapter 3 I will show that it is rational in several 2×2 games for a player to depart from its best state (4), because in these games it would do worse if the other player departed first. On the question of who should move from states, see Hamilton and Slutsky (1993).

not terminate moves at the start. In other words, if *C* moved initially, *R* would then be blocked. Hence, blockage occurs at two states when *C* starts the move–countermove process, whereas it occurs only once when *R* has the first move.

The fact that the rational choice depends on which player has the first move – (3,3) is rational if *R* starts, (2,4) if *C* starts – leads to a conflict over what outcome will be selected when the process starts at (2,4). However, because it is not rational for *C* to move from the initial state, *R*'s move takes precedence, according to rule 6, and overrides *C*'s decision to stay. Consequently, when the initial state is (2,4), the result will be the following

Outcome: (3,3).

2 **Initial state** (4,2). The progressions, survivors, stoppages, other blockages, and outcome from this state are as follows:

$$\begin{array}{lccccccccc} & R & & C & & R & & C & & \\ R\text{ starts:} & \underline{(4,2)} & \rightarrow| & (3,3) & \rightarrow & (1,1) & \rightarrow & (2,4) & \rightarrow| & (4,2) \\ \text{Survivor:} & (4,2) & & (2,4) & & (2,4) & & (2,4) & & \end{array}$$

$$\begin{array}{lccccccccc} & C & & R & & C & & R & & \\ C\text{ starts:} & \underline{(4,2)} & \rightarrow|\text{c} & (2,4) & \rightarrow & (1,1) & \rightarrow & (3,3) & \rightarrow & (4,2) \\ \text{Survivor:} & (4,2) & & (4,2) & & (4,2) & & (4,2) & & \end{array}$$

Outcome: (4,2).

Clearly, when (4,2) is the initial state, there is no conflict between *R* and *C* about staying there. Yet while neither player has an incentive to move from (4,2), each player's reasons for stoppage are different. If *R* starts, there is blockage at the start, whereas if *C* starts, there will be cycling back to (4,2). But because cycling is no better for *C* than not moving, *C* will stay at (4,2) according to rule 5, which I indicate by "c" (for cycling) following the arrow at (4,2). This might be interpreted as a special kind of blockage. In any event, there will be a consensus on the part of both players for staying at (4,2).

3 **Initial state** (3,3). The progressions, survivors, stoppages, other blockages, and outcome from this state are as follows:

$$\begin{array}{lccccccccc} & R & & C & & R & & C & & \\ R\text{ starts:} & \underline{(3,3)} & \rightarrow| & (4,2) & \rightarrow & (2,4) & \rightarrow & (1,1) & \rightarrow & (3,3) \\ \text{Survivor:} & (3,3) & & (3,3) & & (3,3) & & (3,3) & & \end{array}$$

C *R* *C* *R*
C starts: (3,3) → (1,1) → (2,4) →| (4,2) →| (3,3)
Survivor: (2,4) (2,4) (2,4) (4,2)

Outcome: (2,4).

As from initial state (2,4), there is a conflict. If *R* starts, (3,3) is the rational choice, but if *C* starts (2,4) is. But because C's move takes precedence over R's staying, the outcome is that which *C* can induce – namely, (2,4).

4 **Initial state** (1,1). The progressions, survivors, stoppages, other blockages, and outcome from this state are as follows:

R *C* *R* *C*
R starts: (1,1) → (2,4) →| (4,2) →| (3,3) →| (1,1)
Survivor: (2,4) (2,4) (4,2) (3,3)

C *R* *C* *R*
C starts: (1,1) → (3,3) →| (4,2) → (2,4) → (1,1)
Survivor: (3,3) (3,3) (2,4) (2,4)

Outcome: Indeterminate – (2,4)/(3,3) – depending on whether *R* or *C* starts.

Unlike the conflicts from initial states (2,4) and (3,3), it is rational for both players to move from initial state (1,1). But, strangely enough, each player would prefer that the other player be *P*1, because

R's initial move induces (2,4), *C*'s preferred state; and
C's initial move induces (3,3), *R*'s preferred state.

Presumably, each player will try to hold out longer at (1,1), hoping that the other will move first. Because neither player's move takes precedence according to the rules of play (in section 5.1 I shall show how "order power" establishes precedence), neither rational choice can be singled out as the outcome. Hence, I classify the state, when play starts at (1,1), as indeterminate – either (2,4) or (3,3) may occur, depending on which player *P*1 is, which I write as (2,4)/(3,3). Because the choice of first mover is not specified exogenously in this situation (i.e., by the rules of play), indeterminacy emerges endogenously – it is a consequence of TOM.

Typically, this kind of indeterminacy characterizes bargaining (Brams, 1990), wherein each player tries to hold off being the first to make concessions. Although both players would benefit at either (2,4) or (3,3) over (1,1), there is greater benefit to each in letting the other player move first.

Note, however, that the state which R most prefers, (4,2), is unattainable from (1,1) – it can occur only if the process starts at (4,2).

To summarize, each of the initial states goes into the following final determinate states, or outcomes – except when there is a conflict, as there is from (1,1), and neither player's move takes precedence, according to rule 6:

$$(2,4) \rightarrow (3,3);\ (4,2) \rightarrow (4,2);\ (3,3) \rightarrow (2,4);\ (1,1) \rightarrow (2,4)/(3,3).$$

I call the outcomes into which each state goes *nonmyopic equilibria* (NMEs), because they are the consequence of both players' looking ahead and making rational calculations of where, from each of the initial states, the move–countermove process will end up.

The move–countermove process can be interpreted as a bargaining process, in which, starting at the initial state, a player can choose not to move (i.e., to accept an offer) or to move (i.e., reject an offer). If a player chooses to move, the other player can then terminate the game by accepting the offer, or continue it by moving to an adjacent offer, which may in turn be accepted or rejected, and so on.

If this process did not stabilize, the initial offer (i.e., the first move) would not be worth making. But every 2×2 game contains at least one NME, because from each initial state there is an outcome (perhaps indeterminate) of the move–countermove process.[20] If this outcome is both determinate and the same from every initial state, then it is the only NME; otherwise, there is more than one NME.

In the offer/counteroffer interpretation, then, there is at least one offer that will always be accepted, so the process always stabilizes. But this may occur at more than one outcome if there is more than one NME, as in game 56. In this situation, there might be a kind of positioning game played over the choice of an initial state, which we shall analyze in terms of an "anticipation game" in chapter 2.

In game 56, there are three different NMEs, which is the maximum

[20] There was no such existence result in the original theory of moves – as first developed in Brams and Wittman (1981) and then extended in a series of articles that are summarized in Brams (1983) – in which the move–countermove process was assumed to terminate only if the player with the next move reached a state that gave it its best payoff (4). Because this point is never reached in 41 of the 78 2×2 strict ordinal games, no predictions of stable outcomes could be made for the majority of the 78 games, based on the original theory. On the other hand, a weaker equilibrium concept, called an "absorbing outcome," was proposed by Brams and Hessel (1982) for these 41 games, but it, like a nonmyopic equilibrium in the original theory, is *ad hoc* and also not consistent with some of the original nonmyopic equilibria in the 37 games that possess them. Extensions of nonmyopic equilibria have been proposed in Kilgour (1984, 1985), Zagare (1984), and Aaftink (1989). Marschak and Selten (1978), who analyze conditions under which states restabilize if there is a departure from them, and Hirshleifer (1985), who examines different protocols of play, have investigated related equilibrium concepts.

number that can occur in a strict ordinal 2×2 game; the minimum, as already noted, is one. Most 2×2 games have either one or two NMEs, as shown in the Appendix, wherein games are classified according to the number of their NMEs. The point I wish to emphasize here is that *where* play starts in a game can matter, which the unique (2,4) Nash equilibrium in game 56, based on the standard theory, masks.[21]

1.5 Interpreting TOM: Samson and Delilah

That the starting point matters is well illustrated by a famous story from the book of Judges in the Hebrew Bible (Old Testament). To retell this story and explicate the seemingly irrational behavior of the protagonist, it is useful to offer some background.[22]

After abetting the flight of the Israelites from Egypt and delivering them into the promised land of Canaan, God became extremely upset by their recalcitrant ways and punished them severely:

> The Israelites again did what was offensive to the Lord, and the Lord delivered them into the hands of the Philistines for forty years. (Judges 13:l)

But a new dawn appears at the birth of Samson, which is attended to by God and whose angel predicts: "He shall be the first to deliver Israel from the Philistines" (Judges 13:5).

After Samson grew up, he quickly manifested carnal desires that were quite ecumenical:

> Once Samson went down to Timnah; and while in Timnah, he noticed a girl among the Philistine women. On his return, he told his father and mother, "I noticed one of the Philistine women in Timhah; please get her for me as a wife." His father and mother said to him, "Is there no one among the daughters of your own kinsmen and among all our people that you must go and take a wife from the uncircumcised Philistines?" (Judges 14:1–3)

Samson, however, was not to be deterred by such remonstrations, though the Bible explains that God was surreptitiously manipulating events:

21 Even in 2×2 games, Nash equilibria in pure strategies may not be unique, as games I shall analyze later will illustrate. To purge games of such ambiguity, various criteria of equilibrium selection, initiated by Selten (1975), have been proposed that single out one equilibrium (Harsanyi, 1977; Harsanyi and Selten, 1988; Güth and Kalkoften, 1989, van Damme, 1991a; Carlsson and van Damme, 1991; Matsui and Matsuyama, 1991). As I suggested in the Introduction, however, there is not necessarily a problem with multiple equilibria, including NMEs, if they are clearly related to the initial conditions of a game. Indeed, when the starting point is known, it *should* be taken into account in making predictions about stable future outcomes.

22 All translations are from *The Prophets* (1978); the description of this story is adapted from Brams (1980, pp. 153–60) with permission.

His father and mother did not realize that this was the Lord's doing: He was seeking a pretext against the Philistines, for the Philistines were ruling over Israel at that time. (Judges 14:3-4)

The woman whom Samson sought indeed pleased Samson, and he took her as his wife. At a feast, Samson posed a riddle that stumped everybody, and the celebrants appealed to Samson's wife for help:

Coax your husband to provide us with the answer to the riddle; else we shall put you and your father's household to the fire; have you invited us here in order to impoverish us? (Judges 14:15)

Samson's wife was distraught and accused her husband of not loving her, even hating her. At first Samson refused to tell his wife the answer to the riddle, but because she "continued to harass him with her tears . . . on the seventh day he told her, because she nagged him so" (Judges 14:17).

Angered by the whole business, Samson "left in a rage for his father's house" (Judges 14:19). Ironically, "Samson's wife then married one of those who had been his wedding companions" (Judges 14:20).

Samson then had second thoughts about abandoning his wife. When told by his wife's father that it was too late to reconsider, Samson flew into a rage and declared: "Now the Philistines can have no claim against me for the harm I shall do them" (Judges 15:3).

After devastating their fields and vineyards, Samson fought a couple of vicious battles with them. Included among the victims in these cruel encounters were Samson's ex-wife and father-in-law, who were burned to death by the Philistines.

Samson came out of these battles with a reputation as a ferocious warrior of inhuman strength. This served him well as judge (leader) of Israel for twenty years. Samson also cemented his reputation as a man of the flesh by his encounters with prostitutes and other dalliances.

This background on Samson's early life, I believe, helps to make perspicuous his reckless, or at least intemperate, behavior in his last and fatal tryst with a woman named Delilah. She was a Philistine with whom Samson fell in love.

Apparently, Samson's love for Delilah was not requited. Rather, Delilah was more receptive to serving as bait for Samson for appropriate recompense. The lords of the Philistines made her a proposition:

Coax him and find out what makes him so strong, and how we can overpower him, tie him up, and make him helpless; and we'll each give you eleven hundred shekels of silver. (Judges 16:5)

After assenting, Delilah asked Samson: "Tell me, what makes you so strong? And how could you be tied up and be made helpless?" (Judges 16:6). Samson replied: "If I were to be tied with seven fresh tendons, that had not been dried, I should become as weak as an ordinary man" (Judges 16:7).

After Delilah bound Samson as he had instructed her, she hid men in the inner room and cried, "Samson, the Philistines are upon you!" (Judges 16:9). Samson's lie quickly became apparent:

Whereat he pulled the tendons apart, as a strand of tow [flax] comes apart at the touch of fire. So the secret of his strength remained unknown.

Then Delilah said to Samson, "Oh, you deceived me; you lied to me! Do tell me how you could be tied up." (Judges 16:9–10)

Twice more Samson gave Delilah erroneous information about the source of his strength, and she became progressively more frustrated by his deception. In exasperation, Delilah exclaimed:

"How can you say you love me, when you don't confide in me? This makes three times that you've deceived me and haven't told me what makes you so strong." Finally, after she had nagged him and pressed him constantly, he was wearied to death and he confided everything to her. (Judges 16:15–17)

The secret, of course, was Samson's long hair. When he told his secret to Delilah, she had his hair shaved off while he slept. The jig was then up when he was awakened:

For he did not know that the Lord had departed him. The Philistines seized him and gouged out his eyes. They brought him down to Gaza and shackled him in bronze fetters, and he became a mill slave in the prison. After his hair was cut off, it began to grow back. (Judges 16:20–2)

Thus is a slow time bomb set ticking. Ineluctably, the climax approaches when Samson is summoned and made an object of derision and sport by the Philistines in a great celebration:

They put him between the pillars. And Samson said to the boy who was leading him by the hand, "Let go of me and let me feel the pillars that the temple rests upon, that I may lean on them." Now the temple was full of men and women; all the lords of the Philistines were there, and there were some three thousand men and women on the roof watching Samson dance. Then Samson called to the Lord, "O Lord god! Please remember me, and give me strength just this once, O God, to take revenge of the Philistines if only for one of my two eyes." (Judges 16:25–8)

Samson, his emasculated strength now restored, avenged his captors in an unprecedented biblical reprisal (by a human being, not God) that sealed both his doom and the Philistines':

He embraced the two middle pillars that the temple rested upon, one with his right arm and one with his left, and leaned against them; Samson cried, "Let me die with the Philistines!" and he pulled with all his might. The temple came crashing down on the lords and on all the people in it. Those who were slain by him as he died outnumbered those who had been slain by him when he lived. (Judges 1:29–30)

There is irony, of course, in this reversal of roles, whereby the victim becomes the vanquisher. I do not suggest, however, that Samson, intrepid warrior that he was, planned for his own mutilation and ridicule only to provide himself with the later opportunity to retaliate massively against the Philistines. Perhaps this was in God's design, as foretold by the angel at Samson's birth. The more explicit reference to God's "seeking a pretext against the Philistines" (Judges 14:4) when Samson married reinforces this view.

To me, however, these auguries smack of insertions probably made for didactic purposes. They are not central to the narrative, which proceeds well enough without these lessons being drawn.

The didactic references present another problem: God's apparent meddling contradicts the free will humans are presumed to have. It is a tenet of game theory that players make their own independent choices, based on the information they possess about the game being played.

Also puzzling is why, if God is playing some kind of undercover game, He should want so much to help the Israelites after the Bible reports that He delivered them into the hands of the Philistines?[23] In short, the signals given by references to God's purpose and control over events are confusing.

By comparison, I think Samson's behavior as both a truculent warrior and an insatiable lover is consistent and credible. On occasion, perhaps, Samson's strength strains credibility, as when he reportedly slays a thousand men with the jawbone of an ass. Other incidents in his life, such as tearing a lion to pieces, also are stupendous feats, but they are really no more than the normal hyperbole one finds in the Bible. Unquestionably, miraculous achievements, whether God-inspired or not, add drama to the stories, but in my opinion a game-theoretic interpretation should not stand or fall on whether they can be explained in everyday terms.

If Samson's immense strength, or its source, seem beyond human capabilities, his passion for women is not so difficult to comprehend. As the story of his adult life makes clear, Samson lusted after several women, and Delilah was not the first to whose blandishments he fell prey. When he earlier caved in to his wife after she badgered him for several days about the riddle, the pattern was set: He could not withhold information if the right woman was around to wheedle it out of him. While Samson

[23] The quick answer, I suppose, is that they were His chosen, if wayward, people, whom He never intended to abandon forever.

		Samson	
		Don't tell secret ($\bar{T}$)	Tell secret (T)
Delilah	Don't nag Samson ($\bar{N}$)	I Delilah unhappy, Samson unforthcoming (2,4) ↓	II Delilah happy Samson forthcoming (4,2)
	Nag Samson (N)	IV Delilah frustrated Samson harassed (1,1) →	III Delilah persuasive Samson reluctant (3,3)

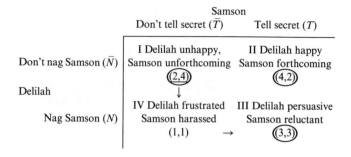

Key: (x,y) = (payoff to Delilah, payoff to Samson)
4 = best; 3 = next best; 2 = next worst; 1 = worst
Nash equilibrium underscored
NMEs circled
Arrows indicate progression of states to NME of (3,3)

Figure 1.2 Samson and Delilah (game 56)

could fight the Philistines like a fiend, he could readily be disarmed by women after whom he hankered.

The payoff matrix of the game I posit that Samson played with Delilah is shown in figure 1.2. Samson's desire having been kindled, Delilah could trade on it either by nagging Samson for the secret of his strength (N) or not nagging him ($\bar{N}$) and hoping it would come out anyway. Samson, in turn, could either tell (T) the secret of his strength or not tell it ($\bar{T}$).[24] Consider the consequences of each pair of strategy choices, starting from the upper left-hand state and moving in a clockwise direction:

I **Delilah unhappy, Samson unforthcoming:** (2,4). The next-worst state for Delilah, because Samson withholds his secret, though she is not frustrated in an unsuccessful attempt to obtain it; the best state for Samson, because he keeps his secret and is not harassed.

II **Delilah happy, Samson forthcoming:** (4,2). The best state for Delilah, because she learns Samson's secret without making a pest of herself; the next-worst state for Samson, because he gives away his secret without good reason.

III **Delilah persuasive, Samson reluctant:** (3,3). The next-best state for both players, because though Delilah would prefer not to nag (if Samson tells) and Samson would prefer not to succumb (if Delilah does not nag), Delilah gets her way when Samson tells; and Samson, under duress, has a respectable reason (i.e., Delilah's nagging) to tell.

IV **Delilah frustrated, Samson harassed:** (1,1). The worst state for both

[24] Although Samson can move from $\bar{T}$ to T, he cannot reasonably move from T to $\bar{T}$ by retracting his secret once it is out, which I call an "infeasible" move in section 1.6.

players, because Samson does not get peace of mind, and Delilah is frustrated in her effort to learn Samson's secret.

The figure 1.2 game is none other than game 56, though I shall consider a plausible reordering of Delilah's preferences shortly.

The game starts at (2,4), when Delilah chooses $\bar{N}$ and Samson chooses $\bar{T}$ during their period of acquaintance. These strategy choices are consistent with Delilah's choosing her dominant strategy, and Samson his best response to this strategy, giving the Nash equilibrium of (2,4) according to the standard theory.

The standard theory offers no explanation of why the players would ever move to a nonequilibrium outcome.[25] But this is precisely what they do. Delilah switches to *N*, putting the players in state (1,1), and Samson responds with *T*, leading to state (3,3), neither of which is a Nash equilibrium.

By contrast, TOM leads to a different prediction when the initial state is (2,4), which I label game 56/I: the configuration of payoffs is game 56, and the state where play commences is I (upper left-hand cell, as shown in figure 1.2). From this state, TOM predicts the outcome to be (3,3). Although (2,4) is also an NME, it does not arise unless play starts in state III – that is, if the game is 56/III, which is not descriptive of the biblical story.[26]

Thus, TOM leads to a unique prediction of (3,3) if play starts in state I, which is the actual outcome of the story. Doubtless, Samson did not anticipate having his eyes gouged out, and being derided as a fool before the Philistines, when he responded to Delilah's nagging by revealing his secret. On the other hand, because he was later able to wreak destruction on thousands of Philistines at the same time that he ended his own humiliation, it seems not unfair to characterize the dénouement of this story as next best for Samson. Perhaps Samson foresaw that revealing his secret to Delilah could create problems, but he almost surely never anticipated that this choice would result in vilification, mutilation, and ultimately death.

Although he surrendered his secret to the treacherous Delilah, Samson apparently never won her love, which seems to be the thing he most

25 To be sure, an extensive-form game could be defined, starting at (2,4), in which other outcomes become the Nash equilibria, including the actual outcome, (3,3), chosen by the players. Thus, it may be unfair to imply that the standard theory fails when a different representation of the game being played might work in the sense of giving the "right" Nash equilibrium. What I seek, however, is not just to explain strategic behavior in the many examples in this book but also to draw general theoretical conclusions about stability in all the 2×2 configurations, based on different rules of play, power balances and imbalances, and information symmetries and asymmetries.

26 Recall that either (3,3) or (2,4) might be the outcome if the initial state were (1,1), the possibility of which I consider in section 5.1.

wanted. In fact, Delilah's decisive argument in wheedling the truth from Samson was that because he did not confide in her, he did not love her. What better way was there for Samson to scotch this contention, and prove his love, than to comply with her request, even if it meant courting not just Delilah but disaster itself?

As for Delilah's preferences, I think it hard to quarrel with the assumption that her two best states were associated with Samson's choice of *T*. I am less sure, however, about the order of preferences she held for her two worst states. It seems to me, contrary to the representation in figure 1.2, that Delilah might have preferred to nag Samson, had he in the end repudiated her, than not nag him. For even though she would have failed to ferret out Samson's secret, Delilah would perhaps have felt less badly after having tried than if she had made no effort at all.

If this is the case, then 2 and 1 for Delilah would be interchanged in the payoff matrix of figure 1.2, and the game configuration would be different (game 47 in the Appendix, with its columns and rows interchanged). But like game 56/I, game 47/I yields (3,3) as the NME when play starts at (1,4), so this reordering of Delilah's preferences would not affect the rational outcome, according to TOM: thinking nonmyopically, Delilah would still switch to *N*, and Samson in turn would switch to *T*.[27]

1.6 Feasible and infeasible moves

So far I have implicitly assumed that players, by switching strategies, can make moves that effect different states in a game; in the end, these moves terminate at some NME. After Samson fell in love with Delilah, for example, she switched from being a seductress to being a nag, and Samson in turn switched – admittedly, reluctantly – from being mum about the secret of his strength to letting it out.

Delilah got her silver, but Samson suffered greatly for succumbing to his desire for Delilah. Nevertheless, Samson seemed to believe, when he made his decision to reveal his secret, that it would lead to a preferred state. For one thing, he detested being nagged by Delilah, as he did by his wife earlier. For another, he had great difficulty resisting women who

27 As I show in Brams (1980, pp. 158–60), (3,3) is also the outcome predicted by the standard theory if Delilah is assumed to move first in 2×4 expansions of games 56 and 47. But the assumption of a specific first mover that renders (3,3) a Nash equilibrium in the 2×4 game is somewhat *ad hoc*. By comparison, the backward-induction analysis of game 56 in section 1.4 demonstrates that, starting at (2,4), it is rational only for *R* (Delilah) to move from (2,4), and similarly from (1,4) in game 47. In other words, TOM enables one to derive which player(s) will move from an initial state instead of assuming, perhaps arbitrarily, that one or the other will play the role of first mover. Hamilton and Slutsky (1993) make the order of moves endogenous in an "extended game."

exploited the love (lust?) he had for them. While Samson's response to Delilah, fueled as it was by sexual attraction, might seem overwhelmingly emotional if not irrational, preferences rooted in emotions are perfectly acceptable from a game-theoretic point of view.

In fact, emotions, as much as more "objective" factors, define a player's preferences.[28] Whatever the basis of a player's preferences, however, they are neither rational nor irrational. Only the choices (of strategies or moves) that players make, foundered on these preferences, can be evaluated in terms of their rationality, based on the rules of play and the rationality rules.

But rational choices may be *infeasible* because, practically speaking, a player cannot make them due to "environmental constraints" (Zagare, 1984, p. 4). For example, suppose that the game between Samson and Delilah in figure 1.2 does not start at (2,4) but instead commences at (3,3). Then TOM predicts that the game will move from (3,3) to (1,1), and thence to (2,4), where moves will cease.

But Samson's initial move from T to $\bar{T}$ makes no sense: once his secret is out, he can hardly retract it, especially in light of the fact that Delilah tested every explanation, true or false, that he gave for his colossal strength. So this move must be ruled out, but not because it is irrational – it induces Delilah subsequently to move from (1,1) to (2,4), which certainly is feasible (she can stop nagging) and best for Samson. Rather, it is incoherent: it contradicts what a reasonable interpretation of strategies in this game would permit. Whereas it is permissible for Samson to move from $\bar{T}$ to T, a reverse switch is hard to entertain in the context of the story, though not necessarily in other interpretations of game 56.

For example, consider the original interpretation of this game given in figure 1.1, in which (3,3) is "Compromise." Starting from this state, one can readily imagine situations in which C will switch from t_2 to t_1, plunging the players into "Disaster," if C believes that this move will drive

28 Howard (1992) and Howard, Bennett, Bryant, and Bradley (1993) develop a "theory of drama," rooted in part in game theory, which makes emotions the means by which players escape the so-called paradoxes of rationality (Howard, 1971). This ingenious twist turns game theory on its head, as it were: rational players find it expedient to react emotionally – even "irrationally" – to the paradoxes with threats, promises, and the like; in the end, they alter their preferences to achieve a dramatic resolution of their conflict and a new self-realization of their situation. Drama theory is compared with TOM in a forum, "Theory of Moves, Game Theory, and Drama Theory," that involves an exchange between Brams (1993b) and Howard (1993). On the strategic role that emotions play in conflict, see Frank (1988). Another source of emotions, like anger and surprise, are the prior beliefs or expectations that players have, and whether or not outcomes conform with them. Beliefs and expectations are incorporated into the structures of games in Gilboa and Schmeidler (1988), where the resulting games are called "information-dependent games," and Geanakopolos, Pearce, and Stacchetti (1989), where they are called "psychological games."

R to switch from s_2 to s_1, yielding "C succeeds," with payoffs of (2,4), as the outcome.

In using TOM to model a strategic situation, the modeler must be sensitive to what strategy changes are feasible and infeasible in this situation. Also, even if moves are rational, the state to which a game moves may, if feasible, be one from which a player cannot move on.

As a case in point, assume C, starting at (3,3) in the figure 1.1 game, switches from t_2 to t_1, changing the state to (1,1). Given that this move is feasible (which I assumed was not the case in the Samson and Delilah story), can R subsequently move on to (2,4)?

I suggest that the answer to this question depends on what interpretation one gives to "Disaster" at (1,1). If this state means nuclear war, nobody may survive its destruction in order to be able to "move on."[29] On the other hand, if C's move to (1,1) creates a severe crisis, like the Cuban missile crisis of 1962 (section 5.4), R may be able to respond, as the Soviet Union did, by withdrawing its nuclear weapons from Cuba, which abated the crisis to the benefit of both the United States and the Soviet Union. Still, the United States was perceived by many analysts as "winning," which is consistent with the upper left-hand (4,2) state in which R is the United States and C is the Soviet Union.[30]

To conclude, I have shown that, given the rules of TOM, the starting point of play may matter. Game 56 has three different NMEs, depending on the initial state, whereas standard game theory singles out only one, which was not the outcome in the Samson and Delilah story. Nevertheless, not all rational moves may be feasible, either because changing strategies, or moving from certain states, violates the *interpretation* of a game.

29 Declaring some moves to be infeasible, or some states to be impossible to move from (i.e., "black holes") for one or both players, is not expressed in the rules of TOM. But such prohibitions can be incorporated into the theory – when applied to specific situations – which may eliminate some NMEs or create new ones.

30 To the best of my knowledge, nobody has proposed game 56 as a model of the Cuban missile crisis, though its payoffs are the same as those in game 57 (Chicken). However, because they are arranged in a different configuration, which I shall analyze in section 5.3, Chicken is an entirely different game.

2 The anticipation problem: there may be no resolution

2.1 Introduction

In chapter 1 I illustrated, using just one game, a startling difference between the predictions of the standard theory (i.e., based on Nash equilibria) and TOM. Whereas TOM predicts one of three possible NMEs in game 56, depending on the initial state, the standard theory picks out only one of these NMEs as *the* (Nash) equilibrium. But if the players anticipate the NMEs in a prior "anticipation game," the one that is also the Nash equilibrium will be chosen, as I shall show later. However, this may not always be the case, which gives rise to an "anticipation problem."

I begin the analysis in this chapter with one of these games, which is a game of *total conflict*: what is the best (4) outcome for one player is worst (1) for the other, and what is next best (3) for one is next worst (2) for the other. This is game 11 and is shown in figure 2.1. To characterize these states verbally, let 4 indicate that a player "wins," 3 that it is "advantaged"; not shown, 2 means that it is "disadvantaged," and 1 that it "loses."

		Column (C)		
		t_1	t_2	
Row (R)	s_1	C advantaged (2,3)	R wins (4,1)	← Dominant strategy
	s_2	C wins (1,4)	R advantaged (3,2)	
		↑ Dominant strategy		

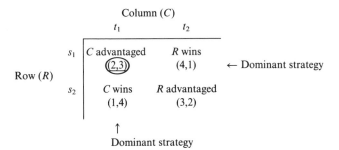

Key: (x,y) = (payoff to R, payoff to C)
4 = best; 3 = next best; 2 = next worst; 1 = worst
Nash equilibrium underscored
NME circled

Figure 2.1 A total-conflict game (game 11)

Observe that the sum of the ranks for the players in each state in game 11 is 5. If these ranks were cardinal utilities, then game 11 would be a *constant-sum game*, wherein the constant is 5. This game would be *zero-sum* if 2.5 were subtracted from each player's utility, so that, for example, the payoffs at (4,1) would be (1.5,–1.5), and the payoffs at (3,2) would be (0.5,–0.5).

In each case, the sum of the payoffs is 0. In fact, a zero-sum game is one in which the (positive) gain of one player exactly matches the (negative) loss of the other player; it is a special case of a constant-sum game.

In constant-sum games, wealth is neither created nor destroyed when players move from one state to another: the gains of one player exactly the match losses of the other. For example, if the state of the game shifts from (4,1) to (2,3) in game 11 because C switches from t_2 to t_1 when R chooses s_1, C gains 2 so-called utiles and R loses 2.[1]

In this sense conflict is *total*, or pure, because there is no possibility for joint gains (or losses). In game 56 (figures 1.1 and 1.2), on the other hand, both players gain at (3,3) *vis-à-vis* (1,1).

Because both players are also better off at (2,4) and (4,2) than (1,1), (1,1) is a *Pareto-inferior* state. The three other states are all *Pareto-optimal*, because there is no other state in game 56 at which both players are better off; moreover, these three states are *Pareto-superior* to (1,1), because they are better for both players. By contrast, all states are Pareto-optimal in game 11, and none is Pareto-superior to any other. As shown in figure 2.1, (2,3) is the dominant-strategy Nash equilibrium and the unique NME in game 11.

Because the framework of TOM is ordinal, I shall not speak of numerical gains or losses in games of total conflict. Rather, in section 2.2 I describe, and interpret in the context of elections, the three qualitatively different configurations of total-conflict games, none of which can be transformed into any of the others by simply interchanging players, their strategies, or both. Two have one NME (this includes game 11), which coincides with the unique Nash equilibrium, and the third has two NMEs but not a Nash equilibrium in pure strategies (mixed strategies in such games will be discussed in sections 2.3 and 2.5).

[1] Unless one admits "interpersonal comparison of utilities," which most game theorists shy away from, these gains and losses are not strictly comparable for the two players. The utilities would be comparable, however, if there were a currency, like money, that the players value in the same way. Then, for example, one could say that the utility that R has for a gain of \$2 is the same as the disutility that C has for a loss of \$2. This, however, is a strong assumption to make. By contrast, TOM avoids such comparisons by putting payoffs in an ordinal framework. Specifically, all that TOM assumes, when play moves from one state to another in a total-conflict game like 11, is that the players do not gain or lose simultaneously.

Total-conflict games are not only interesting in their own right but also are of historical importance. Two-person zero-sum games were the first class of games for which a definitive solution was found, based on the so-called Minimax Theorem (von Neumann, 1928), which I shall comment on in section 2.5. Also, virtually all parlor games, from chess to poker, are zero-sum.

For reasons I shall describe in section 2.2, elections are probably the best examples of total-conflict games in politics. I shall use the three distinct 2×2 total-conflict games to model elections with (i) a strong incumbent, (ii) a strong challenger, and (iii) two more-or-less equal challengers. Two of these games have one NME, whereas the third (game 44) contains two, which I shall derive in section 2.3 using backward induction.

Unfortunately, TOM provides no basis for choosing one or the other of the NMEs in game 44, even in an "anticipation game" in which the players can anticipate the NMEs in game 44 and select strategies beforehand that define an initial state. I contrast this situation with the easier task that players have in the anticipation game of game 56 (analyzed in chapter 1), though the latter has more NMEs (three) than does the anticipation game of 44 (two).

The anticipation problem in game 44 is not peculiar to games of total conflict but also occurs in five different 2×2 games of partial conflict. In these games, like game 44, anticipating the two NMEs does not imply the selection of one (as it does in all other games with two or more NMEs) if the players can make prior strategy choices.

In section 2.4, I use one of the five partial-conflict games in which there is an anticipation problem to model Pharaoh's pursuit of the Israelites, and his confrontation with Moses and God, in the Hebrew Bible. I demonstrate that the theoretical resolution that TOM provides agrees with the actual one reported in the Bible.

In section 2.5 I discuss different approaches to the anticipation problem in literature, showing how William Faulkner, better than Edgar Allan Poe or Sir Arthur Conan Doyle, understood its nature in his fiction. I suggest, nevertheless, that this problem probably does not crop up in most real-life games.

2.2 Three different election games

With the possible exception of sports, politics would appear to be the real-life arena in which games of total conflict are most common. Especially when the stakes are high, as they often are in international politics, the desire to win would seem to be strongest.

But exactly the opposite is probably true. In war, the meaning of "winning" has become increasingly ambiguous; indeed, there may be no winners. Consider some of the major wars fought since World War II: the Korean war (1950–3), the Vietnam war (1965–73), the Iran–Iraq war (1980–7), and the six wars between Arabs and Israelis (1948, 1956, 1967, 1969–70, 1973, and 1982–5). The first two wars ended indecisively in 1953 and 1973 in partition of the countries, though Vietnam was united in 1975 when the North successfully invaded the South after American forces had withdrawn.

The Israelis won militarily in all the wars they fought, but their "victories" in the Yom Kippur war (1973) and the Lebanon war (1982–5) were costly and did not terminate hostilities with their enemies, except for Egypt after the Camp David accords and peace treaty in 1978–9. Even the triumph of the U.S.-led coalition that smashed Iraqi military forces in Kuwait in 1991 in the Persian Gulf war has a hollow ring today, with Saddam Hussein still in power at this writing (April 1993).

To be sure, the allied victories in World War I and World War II were decisive, and the victors had a palpable feeling of success. But the aftermath of World War I, especially the harsh 1919 Treaty of Versailles that brought deprivation to Germany but left her intact, sowed the seeds of World War II; and the aftermath of World War II, especially the 1945 Yalta agreement that gave the Soviet Union a more or less free hand in Eastern Europe, led to more than forty years of conflict between the United States and the Soviet Union. Thus, the fruits of victory after each world war were short-lived for the victors.

The superpower arms race exacerbated the economies of both superpowers, most severely that of the Soviet Union. The Soviet economic collapse, and the concomitant breakup of the Soviet Union into a commonwealth in 1991, hardly signaled a great triumph for the United States. Indeed, as the possibility of a nuclear confrontation between the former superpowers has faded – whose potentially devastating consequences are perhaps the best example of a no-win situation – a possibly destructive trade war among the United States, the European Community, and Japan looms (see section 7.6), which again portends no winners.

"Winning," in short, is suffused with ambiguity. War itself, at least on an international scale, seems increasingly an anachronism in the world today (Mueller, 1989). Of course, a game that ends in stalemate may be zero-sum, but most of the foregoing examples bespeak a mutual loss to the players as a result of their conflict.

Elections in democratic countries are the major exception to the decidedly nonzero-sum character of most politics. When one side wins, the other side (or sides) lose. True, in legislative elections the matter may not be so clear-cut when, for example, no party in a parliament wins a majority of seats, or can put together a coalition that controls a majority.

I consider here two-candidate (or two-party) elections for a single office, such as for senator or president. Assume that each candidate may choose either a risky (R) or nonrisky ($\bar{R}$) strategy. I next describe three scenarios, each modeled by a different total-conflict game, and give the NMEs in each. I illustrate the scenarios with examples from all the presidential campaigns since 1960 (except for 1992, which seems more a three-person game) and then focus on the difficulty of choosing a strategy in the third scenario.

1 Strong incumbent: one NME at $R\bar{R}$. The players in this game, 25, are a challenger (C) and an incumbent or office-holder (O), as shown in figure 2.2.[2] Because O is "strong," I assume that its choice of $\bar{R}$ leads to its two best payoffs (3 and 4) – that is, it is always better off playing it safe. Necessarily, its choice of R leads to its two worst payoffs (1 and 2).

Whether O chooses R or $\bar{R}$ in this game, however, it does better when C chooses the same strategy. The reason, I assume, is that O's incumbency is always an advantage in a situation in which the players choose the same strategies (more on the meaning of these strategies shortly).

Because the game is one of total conflict, C's preferences are the mirror image of O's. As shown in figure 2.2, this game has a unique NME, (2,3), when C chooses R and O chooses $\bar{R}$: players will either move to this state or, if it is the initial state, not depart from it. The resulting outcome, in terms of comparative ranks, favors O.[3]

[2] Game 25 in figure 2.2, like other games illustrated throughout this book, appears in a different form in the Appendix, but it is structurally the same. In this case, the row player in figure 2.2 is equivalent to the column player in the Appendix representation, which can be seen from the fact that the row player's two medium states (3 and 2) are associated with its first strategy, just as the column player's two medium states are associated with its first strategy in the Appendix. Also, departures from (3,2) and (2,3) by the row player in figure 2.2 lead to (4,1) and (1,4), which is equivalent to departures from (2,3) and (3,2) by the column player in the Appendix representation that lead to (1,4) and (4,1), respectively.

[3] The outcome (2,3) is also the unique Nash equilibrium in game 25, as is (3,2) in game 11 that models scenario 2. Because these are games of total conflict, the Nash equilibrium in each is a *saddlepoint*, which is an entry in a payoff matrix at which the row player receives the minimum payoff in its row and the maximum payoff in its column, or "minimax" for short (I discuss the Minimax Theorem in sections 2.3 and 2.5). Thus, in game 25, (2,3) gives Row the minimum payoff associated with its R strategy ($2<3$) and the maximum payoff associated with O's $\bar{R}$ strategy ($2>1$). The etymology of "saddlepoint" is as follows: The surface of a saddle curves upward in one direction from the center (i.e., the line of motion of the horse), corresponding to a row minimum, and downward in the other direction, corresponding to a column maximum. Hence, an outcome in a 2×2 game which is simultaneously the minimum of a row and the maximum of a column is analogous to the center of a saddle-shaped surface. (However, the matrix entries of an ordinal game do not possess the property of continuity that characterizes a smooth surface.) Game 44 in figure 2.2 has no saddlepoint, so there is no state at which either the row player ($C1$) cannot do

1 Strong incumbent: game 25

		Office-holder (*O*) *R*	Office-holder (*O*) $\bar{R}$
Challenger (*C*)	Risk (*R*)	(3,2)	(2,3)
	Don't risk ($\bar{R}$)	(4,1)	(1,4)

2 Strong challenger: game 11

		Office-holder (*O*) *R*	Office-holder (*O*) $\bar{R}$
Challenger (*C*)	Risk (*R*)	(3,2)	(4,1)
	Don't risk ($\bar{R}$)	(1,4)	(2,3)

3 Two challengers: game 44

		Challenger 2 (*C*2) *R*	Challenger 2 (*C*2) $\bar{R}$
Challenger 1 (*C*1)	Risk (*R*)	(4,1)	(1,4)
	Don't risk ($\bar{R}$)	(2,3)	(3,2)

Key: (*x*,*y*) = (payoff to *C* or *C*1, payoff to *O* or *C*2)
4 = best; 3 = next best; 2 = next worst; 1 = worst
Nash equilibria underscored
NMEs circled

Figure 2.2 Three total-conflict election games (games 25, 11, and 44)

What does it mean to say that a strategy is "risky"? In general, choosing *R* means selecting a campaign strategy with less predictable consequences than choosing $\bar{R}$. For example, to mount a negative campaign is usually risky precisely because it may either succeed brilliantly or backfire badly.

I do not assume that "4" means that a candidate will always win, "3" that it is favored to win, and so on; these ranks are assumed only to be relative. For example, if *C* has virtually no chance of winning, a state in

better by switching to the strategy that gives it a maximum in a column, or the column player (*C*2) cannot do better by switching to the strategy that gives it a maximum in its row. The absence of a saddlepoint in game 44 is why preferences in this game cycle, with one player having an incentive to depart from every state in a clockwise direction (cyclicity in the 2×2 games is formally analyzed in section 4.2).

which it receives a payoff of 4 means simply that it does better than in any other state of the game.

The presidential campaigns of 1964 and 1972 illustrate the relativistic nature of the ranks in this scenario. Neither Barry Goldwater in 1964 nor George McGovern in 1972 had any serious chance of unseating each of their incumbent opponents, Lyndon Johnson or Richard Nixon. Although Walter Mondale's prospects in 1984 against another strong incumbent, Ronald Reagan, looked almost as dismal, the race was in fact fairly close until the end, which Reagan won in a landslide.

In relating the NME of (2,3) in game 25 to the aforementioned campaigns, it seems fair to say that the incumbent in each case chose $\bar{R}$. Johnson portrayed himself as the experienced politician, who would continue the legacy of John Kennedy and not escalate the Vietnam conflict; Nixon hardly campaigned, pursuing an aloof "Rose Garden" strategy from the White House; and Reagan promised more good times, following the economy's rapid recovery from a severe recession in the early 1980s.

By contrast, each of the challengers, out of necessity if not desperation, attacked vigorously and took major chances. However, except for some obvious blunders each made (Goldwater appeared trigger-happy about the use of nuclear weapons; McGovern stuck too long to his first vice-presidential choice, Thomas Eagleton, after Eagleton's hospitalization for depression was revealed; and Mondale said he would raise taxes), it is not clear that any of the challengers could have done appreciably better by being less bold.

I therefore believe that the joint choice of $R\bar{R}$ by C and O, respectively, in game 25 plausibly models the actual choices of the candidates in the three presidential campaigns I have discussed. I next turn to a scenario in which C and O switch places, with C now having the upper hand.

2 Strong challenger: one NME at *RR*. Assume C is "strong": its choice of R leads to its two best payoffs (3 and 4) – that is, it is always better off taking risks. Necessarily, its choice of $\bar{R}$ leads to its two worst payoffs (1 and 2). Whether it chooses R or $\bar{R}$, C does better when O chooses $\bar{R}$, because $\bar{R}$ reminds voters of the shakiness of O's incumbency.

As shown in figure 2.2, this game is 11 – the same game as that given in figure 2.1 (with the two row and the two column strategies interchanged) – and has a unique NME of (3,2) when both players choose R. This outcome, in terms of comparative ranks, favors C.

An example of a presidential campaign that mirrors game 11 is Ronald Reagan's successful challenge of Jimmy Carter in 1980, who was saddled with raging inflation and American hostages in Iran (the hostage crisis is modeled in section 6.4). As a challenger himself four years earlier, Carter had defeated Gerald Ford; in fact, this campaign offers a second example of scenario 2. Ford, who had become president in 1974 after Richard Nixon's resignation, was certainly not helped in his 1976 campaign by his pardon of Nixon.

In both 1976 and 1980, the challengers selected similar strategies, strongly disassociating themselves from the so-called Washington establishment. (This might not seem exactly risky today, but a Washington outsider prior to 1976 had not won the presidency since Dwight Eisenhower in 1952, and he was already extremely well known.) Both Carter and Reagan were ex-governors, and hence credible in this role, but by 1980 Carter could no longer cast himself as an outsider. Less still could Gerald Ford in 1976, who had been a long-time member of the House of Representatives and then Nixon's vice president before Nixon resigned.

Nevertheless, perhaps realizing their tenuous reelection prospects, both candidates selected what might be considered risky strategies (though a case can be made that neither's tactics were rash). Ford strongly defended his presidential pardon of Nixon, though it was highly unpopular, and Carter tried to ignore his challenger in the beginning, refusing to participate in the first presidential debate (ostensibly because it included the third-party nominee, John Anderson). Ford, in the end, lost in a close race in 1976, but Carter was resoundingly defeated in 1980.

So was George Bush in 1992, who arguably played it safe by echoing his 1988 campaign (on this campaign, see scenario 3), contrary to the prediction of game 11. But the presence of Ross Perot, who garnered 19 percent of the popular vote in the 1992 race that also included Bill Clinton, makes modeling this race as just a two-person game problematic.

It seems clear in retrospect that Carter could not have won under any foreseeable circumstances, though what might have constituted for him a "better" R is far from apparent. Perhaps a more coherent and decisive strategy for dealing with the energy crisis and the resulting inflation in the late 1970s, instead of worrying about the "malaise" of the country during his presidency, would have led to a stronger showing.

Ford, on the other hand, conceivably could have reversed his loss, but if game 11 is an accurate representation of the 1976 campaign, his "better" R is not clear. Possibly bolder measures for dealing with inflation, instead of WIN ("whip inflation now") buttons, would have helped.

In sum, the joint choice of RR by C and O in game 11 reflects, in my opinion, the choices of the challengers and, more controversially, the choices of the incumbents in the 1976 and 1980 presidential campaigns. I now turn to the third scenario, in which the candidates are no longer a distinguishable challenger and incumbent. In this game, unlike games 25 and 11, an outcome different from the Nash equilibrium (which coincides with the NMEs in games 25 and 11) was selected.

3 Two challengers: two NMEs at $\bar{R}R$ and $R\bar{R}$. The players in game 44 are challenger 1 (*C*1) and challenger 2 (*C*2), as shown in figure 2.2. *C*1's choice of *R* results either in its best (4) or its worst (1) payoff, making it consonant with the idea of risk. Necessarily, its choice of $\bar{R}$ gives it intermediate payoffs (3 and 2). *C*2 does not have the same choices, obtaining either 3 or 1 when it chooses *R* and 4 or 2 when it chooses $\bar{R}$. Because one player is always motivated to depart from each state in a clockwise direction in this game, preferences cycle.

Both (2,3) and (3,2), associated with *C*1's choice of $\bar{R}$, are NMEs. As I shall show in section 2.3, if the player receiving a payoff of 2 should depart from either state, the process will cycle clockwise back to this state, whereas there is blockage at every move should the player receiving 3 depart initially. Hence, neither player will depart from these states. On the other hand, if (4,1) is the initial state, it will go into (3,2), whereas (1,4) will go into (2,3), so the players suffering their worst states can at least induce their next-worst states.

The existence of two NMEs in game 44 makes the initial state crucial in determining which state will be selected as the outcome. This selection problem is illustrated by the presidential campaigns of 1960, 1968, and 1988, in which there was no incumbent seeking reelection. To relate game 44 to each of these campaigns, I begin by indicating which candidate in each seemed to be *C*1, who had to choose between *R*, with possible payoffs of 4 and 1, and $\bar{R}$, with possible payoffs of 3 and 2.

I believe that Richard Nixon in 1960, the vice president in the previous Republican administration, and Hubert Humphrey in 1968, the vice president in the previous Democratic administration, were *C*1 in these two campaigns, and each, acting somewhat timidly, chose $\bar{R}$ as its strategy. (To be sure, George Wallace was a third-party candidate in 1968, though he was not as significant, as measured by his popular vote of 14 percent, as Ross Perot in 1992, though Wallace was the plurality winner in five states and Perot in none.) Despite Nixon's last-minute accommodation with Nelson Rockefeller to keep Rockefeller from contesting the Republican nomination (Rockefeller was the liberal governor of New York and posed the only serious challenge to Nixon's nomination), Nixon's campaign echoed the policies of the Eisenhower administration. Likewise, Humphrey was a captive of the Johnson administration, though he tried hard to stake out a somewhat different position on the Vietnam war than that which Lyndon Johnson had espoused.

As for their opponents, John Kennedy struck out boldly in 1960, promising to "get the country moving again." It is less obvious that the "new" Nixon in 1968, as $C2$, chose R, but he did trumpet that he had a "plan" (never spelled out) that could end the Vietnam war promptly.

The prospect of escape from the status quo attracted enough voters to put Kennedy in 1960, and Nixon in 1968, over the top, but just barely. Both eked out victories with margins of less than 1 percent in the popular vote, though both won larger majorities in the Electoral College. Their victories, based on choosing R, are consistent with the prediction of (2,3) in game 44, which favors $C2$ (Kennedy in 1960 and Nixon in 1968) when $C1$ (Nixon and Humphrey) chooses $\bar{R}$.

In 1988 this outcome was repeated, but now with a reversal of roles. It was the Democratic nominee, Michael Dukakis, who assumed the mantle of the aforementioned vice presidents (i.e., was $C1$). Because of his enormous early lead in the polls, Dukakis shunned making a provocative challenge to the Republican nominee, George Bush, and chose instead a pallid $\bar{R}$. This put Bush in a position, while hewing to the policies of the Reagan administration but distancing himself from its most embarrassing moment ("Irangate," involving, among other things, an unsuccessful bribery attempt to free American hostages in Iran), to mount a strong negative campaign against Dukakis, which evoked charges of racism. Nonetheless, it was an extremely effective campaign and enabled Bush to overwhelm Dukakis at the polls, again mirroring the (2,3) NME in game 44.

By contrast, there is no Nash equilibrium in "pure" strategies in game 44; I discuss its mixed-strategy equilibrium in section 2.5. Thus, TOM makes an unequivocal "point" prediction, depending on where play commences, whereas the standard theory does not. This fact suggests that positioning before a game, in what may be called its preplay phase, matters, though the extent to which this is possible may be circumscribed by the stances on policy issues that the candidates have taken in the past.

My attempt to relate the eight presidential campaigns from 1960 to 1988 to the NMEs in the three total-conflict games does not, in any rigorous sense, provide a test of TOM, even with respect to these recent elections. Rather, it offers *prima facie* evidence that these different total-conflict games may model, albeit in a highly simplified way, different campaign scenarios that actually occur in elections.

Because the first two games (25 and 11) have unique NMEs, the outcome does not depend on where play starts. But in game 44, with two NMEs, the situation is dependent on the initial state, which raises the question of how this state is determined. (So far I have assumed that it is given.) In section 2.3, I alter this assumption and suppose that players can select strategies in an "anticipation game," which defines an initial state and, therefore, an outcome.

2.3 Anticipation games

Consider game 44 in figure 2.2. Backward induction, as described in section 1.4, leads to the following four outcomes when play commences at each of the initial states:

1 **Initial state** (4,1)

	R		*C*		*R*		*C*		
R starts:	(4,1)	→|	(2,3)	→|	(3,2)	→|	(1,4)	→|	(4,1)
Survivor:	(4,1)		(2,3)		(3,2)		(1,4)		
	C		*R*		*C*		*R*		
C starts:	(4,1)	→	(1,4)	→	(3,2)	→|	(2,3)	→	(4,1)
Survivor:	(3,2)		(3,2)		(3,2)		(4,1)		

Outcome: (3,2). *C*'s move induces (3,2), which takes precedence over *R*'s staying at (4,1).

2 **Initial state** (1,4).

	R		*C*		*R*		*C*		
R starts:	(1,4)	→	(3,2)	→	(2,3)	→|	(4,1)	→	(1,4)
Survivor:	(2,3)		(2,3)		(2,3)		(1,4)		
	C		*R*		*C*		*R*		
C starts:	(1,4)	→|	(4,1)	→|	(2,3)	→|	(3,2)	→|	(1,4)
Survivor:	(1,4)		(4,1)		(2,3)		(3,2)		

Outcome: (2,3). *R*'s move induces (2,3), which takes precedence over *C*'s staying at (1,4).

3 **Initial state** (3,2).

	R		*C*		*R*		*C*		
R starts:	(3,2)	→|	(1,4)	→|	(4,1)	→|	(2,3)	→|	(3,2)
Survivor:	(3,2)		(1,4)		(4,1)		(2,3)		
	C		*R*		*C*		*R*		
C starts:	(3,2)	→|c	(2,3)	→	(4,1)	→	(1,4)	→	(3,2)
Survivor:	(3,2)		(3,2)		(3,2)		(3,2)		

Outcome: (3,2). There is a consensus on staying at (3,2), with constant blockage for *R*, and cycling (c) to (3,2) if *C* initiates a move.

4 **Initial state** (2,3).

	R		*C*		*R*		*C*		
R starts:	(2,3)	→|c	(4,1)	→	(1,4)	→	(3,2)	→	(2,3)
Survivor:	(2,3)		(2,3)		(2,3)		(2,3)		
	C		*R*		*C*		*R*		
C starts:	(2,3)	→|	(3,2)	→|	(1,4)	→|	(4,1)	→|	(2,3)
Survivor:	(2,3)		(3,2)		(1,4)		(4,1)		

Outcome: (3,2). There is a consensus on staying at (2,3), with constant blockage for *C*, and cycling (c) to (2,3) if *R* initiates a move.

To summarize, from initial states (4,1) and (1,4) the pattern is:

- whereas one player will stay, the other player will move;
- because moving takes precedence over staying according to rule 6 (section 1.4), the process will move;
- the moves – (4,1) → (3,2) and (1,4) → (2,3) – give two NMEs, with the player who moves going from its worst (1) to its next-worst (2) state.

By comparison, from initial states (3,2) and (2,3) there is blockage at the start, but the reason depends on the player:

- for one player – *R* at (3,2) and *C* at (2,3) – not only the initial move, but all later moves as well, lead to an inferior outcome for the player who moves and are, therefore, blocked;
- for the other player – *C* at (3,2) and *R* at (2,3) – a move results in cycling back to the initial state, so, by rule 6, it will not be made;
- because the players will stay at (3,2) and (2,3), these states go into themselves – (3,2) → (3,2) and (2,3) → (2,3) – resulting in the same NMEs as those induced by moves from (4,1) and (1,4), respectively.

Game 44 with two NMEs, and game 56 (chapter 1) with three NMEs, are interesting to compare. In figure 2.3, I indicate the outcomes into which each state goes in brackets, just below the original states in parentheses. The bracketed payoffs define what I call the *anticipation game* (AG), whose payoffs are simply the NMEs into which each original state goes.

Suppose that the players, contrary to rule 1 of TOM, do not start at some initial state. Rather, assume they choose strategies in AG in order to select a state, which they anticipate, based on TOM, will lead into a particular outcome (perhaps indeterminate, as in game 56).

If players choose strategies in AG rather than start from states in the original game, then one can apply the standard theory to AG. Thus, in the

Game 56

		Column (C) t_1	t_2	
Row (R)	s_1	(2,4) [3,3]	(4,2) [4,2]	← Dominant strategy
	s_2	(1,1) [2,4]/[3,3]	(3,3) [2,4]	

Game 44

		Column (C) t_1	t_2
Row (R)	s_1	(4,1) [3,2]	(1,4) [2,3]
	s_2	(2,3) [2,3]	(3,2) [3,2]

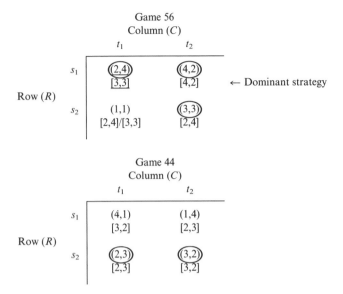

Key: (x,y) = (payoff to R, payoff to C)
$[x,y]$ = [payoff to R, payoff to C] in anticipation game (AG)
4 = best; 3 = next best; 2 = next worst; 1 = worst
Nash equilibrium in game 56 and its AG underscored
NMEs circled

Figure 2.3 Original and anticipation games (games 56 and 44)

AG of game 56 R has a dominant strategy of s_1: whether C chooses t_1 or t_2, R receives at least as high and sometimes a higher payoff by choosing s_1 rather than s_2. Given that C believes R will choose its dominant strategy, C, preferring (3,3) to (4,2), will choose t_1.

In fact, (3,3) is the unique Nash equilibrium in the AG of game 56. Hence, if an initial state is not given (as TOM assumes), but instead the players (i) can choose strategies in a preplay phase and (ii) can anticipate the NME that that state will go into, (3,3) is the NME they will select in the AG.[4] Recall that this is the actual outcome of the Samson and Delilah game (section 1.5), whereas (2,4) is the Nash equilibrium in the original game.

In both game 44 and in its AG, there is no Nash equilibrium in pure strategies (i.e., a saddlepoint). Because one player always has an incentive to depart in a clockwise direction from every state in these games,

[4] Alternatively, applying TOM to the AG of game 56 gives (3,3) as the state that *every* state of the AG goes into – whether the indeterminate state in game 56, (1,1), is assumed to go into (2,4) or (3,3) in the AG – so anticipating the AG itself also gives (3,3) as the outcome.

preferences cycle. Moreover, applying TOM to the AG of game 44 simply reproduces the AG, so no amount of anticipation resolves the problem of choosing a "best" pure strategy. Put another way, neither player has a dominant strategy in either game 44 or its AG, the AG of the AG, and so on, because the AGs simply reproduce themselves.

On the other hand, if the payoffs in these games were cardinal utilities, then the mixed-strategy Nash equilibrium would be for each player randomly to select each of its pure strategies with probability 0.5. This 50–50 mixture is in equilibrium because, given its choice by both players, neither player can do better by switching its strategy.[5]

But payoffs are not cardinal in TOM, so this application of the Minimax Theorem – which ensures that each player can guarantee itself a certain expected value in a two-person constant-sum game (see section 2.5) – to game 44 and its AG are somewhat beside the point. I prefer to say that, anticipating the NMEs in game 44, TOM neither prescribes nor predicts the strategy choices of the players, because the strategies of each player are undominated in its AG. That is, which strategy leads to a better outcome depends on the (undominated) choice of the other player, so it is impossible to anticipate the choice of one or the other, based on rational calculations. Consequently, even though every 2×2 strict ordinal game has at least one NME, it may not be possible to determine which one will be selected, even given that a player can choose strategies in its AG in advance.[6]

This indefiniteness occurs not only in game 44. There are also five 2×2 games of partial conflict in which preferences cycle in both the original game and its AG.[7] Like game 44, these games (42, 43, 45, 46, and 47 in the Appendix) all contain two NMEs, whereas there are three games in which preferences cycle in the original game but not its AG. The latter games – 29, 30, and 31 – all contain only one NME (see Appendix).

Thus, a necessary and sufficient condition for there to be cycling in an AG is that there be cycling in its original game and that it contain two

5 To see this, assume that the column player in game 44 chooses t_1 and t_2 at random, each with probability 0.5. If the row player chooses s_1, it obtains the average of its payoffs of 4 and 1 in the first row, or 2.5. If it chooses s_2, it obtains the average of its payoffs of 2 and 3 in the second row, or also 2.5. Thus, the choice of the 50–50 mixture guarantees an expected payoff of 2.5 for the other player, no matter which pure strategy, or mixture of pure strategies, the other player chooses. I will say more about the interpretation of mixed strategies, which may be mixtures other than 50–50, in section 2.5.

6 If a 2×2 game has only one NME – into which every state of the original game goes – then this is obviously the only outcome that can be chosen in its AG. Thus, the only outcomes in the AGs of games 25 and 11 in figure 2.2 are their NMEs of (2,3) and (3,2), respectively.

7 In section 4.2 I call these games "strongly cyclic": from every state, one player can do immediately better by moving in either a clockwise or a counterclockwise direction (but not both, for reasons given in section 4.2).

NMEs. Although the six aforementioned games in which this occurs would appear to be paragons of instability (or chaos), in reality I believe this is probably not the case. For reasons given in section 1.2, players do not usually start by choosing strategies but, instead, start in states, which terminate at some outcome. (That it may be indeterminate I consider later.)

To illustrate this point, I will use one of the five partial-conflict games with a cyclical AG (game 42) to model a story from the book of Exodus in the Hebrew Bible. This story illustrates how the ambiguity of an AG that makes no definite prediction is resolved – provided play starts in a state that goes into a single NME.

2.4 The pursuit of the Israelites

After Moses, with God at his side, visits ten devastating plagues on Egypt, Pharaoh agrees to let the Israelites flee from their bondage. But did Pharaoh make up his own mind, or was he simply a puppet? Whether Pharaoh was a genuine player at the time of the plagues bears on the question of his status as a player in the next game he played, which is the game I shall focus on here.

Before the plagues, God revealed to Moses something of His grand design against Pharaoh:[8]

> You shall soon see what I will do to Pharaoh: he shall let them [the Israelites] go because of a greater might; indeed, because of a greater might he shall drive them from his land. (Exodus 6:1)

More ominously, God said He would

> harden Pharaoh's heart, that I may multiply My signs and marvels in the land of Egypt. When Pharaoh does not heed you, I will lay My hand upon Egypt and deliver My ranks, My people the Israelites, from the land of Egypt with extraordinary chastisements. (Exodus 7:4–5)

In fact, by rendering Pharaoh obstinate in the face of each more ghastly plague, God both magnifies and glorifies His achievements, which seems to have been precisely His intention.

But in making Pharaoh "stubborn" (Exodus 7:14), God seems to rob him of his free will and his authenticity as a *player* – that is, someone capable of making his own decisions. Nevertheless, Pharaoh eventually reached the point, after the tenth plague that brought death to all the first-born Egyptians (including Pharaoh's), in which dogged defiance

[8] All translations are from *The Torah: The Five Books of Moses* (2nd edn, 1967); the analysis of this story is adapted from Brams (1980, pp. 84–94) with permission.

		God/Moses		
		Help Israelites (H)		Don't help Israelites ($\bar{H}$)
Pharaoh	Pursue (P)	I New confrontation ((2,4)) [2,4]		II Help unforthcoming (4,1) [3,2]
		↑		
	Don't pursue ($\bar{P}$)	IV Help given unnecessarily ((3,2)) [3,2]	←	III Unaided flight successful (1,3) [2,4]

Key: (x,y) = (payoff to Pharaoh, payoff to God/Moses)
$[x,y]$ = [payoff to Pharaoh, payoff to God/Moses] in anticipation game (AG)
4 = best; 3 = next best; 2 = next worst; 1 = worst
NMEs circled
Arrows indicate progression of states to NME of (2,4)

Figure 2.4 The pursuit of the Israelites (game 42)

seemed futile, or at least no better than suicide in installments. Pharaoh then asserted his independence from God's control, making the prudent calculation that it was in his interest to let the Israelites go.

Yet second thoughts soon set in:

When the king of Egypt was told that the people had fled, Pharaoh and his courtiers had a change of heart about the people and said, "What is this we have done, releasing Israel from our service?" (Exodus 14:5)

Now Pharaoh must decide whether to pursue or not pursue the Israelites, which could provoke a new confrontation with God/Moses, whom I interpret as one player. In turn, God/Moses must decide whether to help or not help the Israelites. The payoff matrix for game 42 is shown in figure 2.4, wherein Pharaoh may either pursue (P) or not pursue ($\bar{P}$) the Israelites, and God/Moses may either help (H) or not help ($\bar{H}$) the Israelites. I justify the payoffs at each state, starting with the upper left-hand state and moving in a clockwise direction, as follows:

I **New confrontation:** (2,4). The next-worst state for Pharaoh, because even though the loss of the Israelite slaves would be catastrophic for Egypt, he risks a great deal in a new confrontation with God/Moses; the best state for God/Moses, because they (especially God) relish the opportunity to display their power, even after the plagues.[9]

[9] As evidence, God says, "I will stiffen Pharaoh's heart and he will pursue them, that I may assert My authority against Pharaoh and all his host; and the Egyptians shall know that I am the Lord" (Exodus 14:4).

II **Help unforthcoming:** (4,1). The best state for Pharaoh, because he sees his victory as an easy one against the unarmed Israelites; the worst state for God/Moses, because they do not come to the aid of the Israelites in distress.

III **Help given unnecessarily:** (3,2). The next-best state for Pharaoh, because he would prefer not to pursue the Israelites when they are protected by God/Moses; the next-worst state for God/Moses, because they would be aiding a people quite capable of fending for themselves (i.e., when not pursued).

IV **Unaided flight successful:** (1,3). The worst state for Pharaoh, because he could have regained the Israelite slaves without opposition; the next-best state for God/Moses, because the Israelites escape, though without benefit of another display of God/Moses; power.

Note that preferences cycle in a counterclockwise direction in this partial-conflict game.[10]

The NMEs into which each state goes are shown in the AG matrix in brackets below the payoffs of the original game in figure 2.4. The original game commences in state III at (1,3), which goes into the NME of (2,4) when Pharaoh switches from $\bar{P}$ to P and God/Moses in turn switches from $\bar{H}$ to H.[11]

But the "rationality" of this outcome seems dubious in view of what actually happened in the pursuit game. As the Egyptians bore down on the Israelites, Moses responded to their cries:

> Have no fear! Stand by, and witness the deliverance which the Lord will work for you today; for the Egyptians whom you see today you will never see again. The Lord will battle for you; you hold your peace! (Exodus 14: 13–14)

Indeed, with the help of God, Moses parted the Red Sea before the Israelites and, as soon as they had safely crossed it, directed that the waters flow back. The entire Egyptian army was swallowed up in these waters.

In retrospect, Pharaoh's tenacious pursuit of the Israelites was an unmitigated disaster, but I would argue that he could not possibly have

[10] Also noteworthy is that by interchanging 3 and 4 in God/Moses' preferences, game 42 becomes game 44 in figure 2.3, making it "close" to the latter total-conflict game. As in every partial-conflict 2 × 2 strict ordinal game, there is at least one Pareto-inferior state. In game 42, it is (1,3), which is worse for both players than (2,4).

[11] As in the Samson and Delilah game (section 1.5), (2,4) is also predicted by the standard theory if Pharaoh is allowed to move first in a 2 × 4 expansion of game 42 in figure 2.4. But as earlier, "stabilizing" (2,4) by showing it to be the Nash equilibrium in the 2 × 4 expansion sidesteps the question of why one player (Pharaoh in this case) would move first. By contrast, TOM shows this move to be rational if (1,3) is the initial state (which it was in the story).

anticipated this outcome when he made his strategy choice.[12] At worst, I believe, Pharaoh might have anticipated something of the order of the plagues, each of which at least provided a warning of worse things to come. By comparison, it is reasonable to suppose that it would have been riskier for Pharaoh to have suffered the loss of the slaves without a fight. Then the Egyptians might have risen up in anger against him, deposing or even killing him in state III, which I argued earlier was Pharaoh's worst.

At the same time, the prospects of recapturing the Israelites did not look inauspicious. As the Egyptians closed in on the petrified Israelites, the Israelites were on the verge of deserting their leader:

> Greatly frightened, the Israelites cried out to the Lord. And they said to Moses, "Was it for want of graves in Egypt that you brought us to die in the wilderness? What have you done to us, taking us out of Egypt? Is this not the very thing we told you in Egypt, saying, 'Let us be, and we will serve the Egyptians, for it is better for us to serve the Egyptians than to die in the wilderness?'" (Exodus 14:10–12)

Furthermore, even after God threw the Egyptians into a panic by locking their chariot wheels as they crossed the parted sea, the Egyptians were not beyond making a clearheaded calculation that they thought might still save them: "Let us flee from the Israelites, for the Lord is fighting for them against Egypt" (Exodus 14:25). But by then it was too late.

In summary, Pharaoh calculated that he had a good chance of intercepting the slaves while in flight. Even if God intervened again on the side of the Israelites, as he had with the series of horrendous plagues, Pharaoh did not anticipate his own destruction, at least not in one fell swoop. Thus, I think it fair to say he was a player, at least partially if not totally in control (Brams, 1980, pp. 93–4, 103–4).

The biblical story I have modeled with game 42 and its AG in figure 2.4, despite the cyclical preferences of its players, did not end in irresolution. The reason, as I suggested earlier, is because the game started in a particular state, not by the players' choosing strategies. Not only did it start in a state, as I think most games do, but there are no infeasible switches (section 1.6). Unlike the Samson and Delilah game (figure 1.2), the players can depart from any state, reverse themselves, and continue back and forth without doing violence to the interpretation I give to their strategies.

2.5 The anticipation problem in literature

There may not always be such a deft resolution to games. As I shall

[12] Remember that game 42 and its AG in figure 2.4 reflect the players' preferences as they saw matters at the time. Thus, Pharaoh, like Samson in game 56 (section 1.5), plans his

illustrate in the case of a novel by William Faulkner, two characters may be forced by circumstance to make apparently random choices in a game of total conflict. Unlike Faulkner, however, not all writers choose to highlight the randomness (meaninglessness?) of human choices in their fiction, preferring instead to emphasize the ability of one character to outwit another.[13]

In *The Adventures of Sherlock Holmes*, Sir Arthur Conan Doyle portrayed this element in the difficult choice that he gave to Sherlock Holmes, pursued by the notorious Moriarty, of whether to get off his train at Dover or at Canterbury, an intermediate stop. In the story, Holmes chooses Canterbury, anticipating that Moriarty will take a special faster train to Dover to try to catch him if he gets off there. Holmes' anticipation is correct, but Morgenstern (1935, p. 174) asks the critical question: "But what if Moriarty had been still more clever, had estimated Holmes' mental abilities better and had foreseen his actions accordingly?"

Morgenstern originally posed this question in his first book (Morgenstern, 1928), which, coincidentally, was published the same year as the proof of the Minimax Theorem (von Neumann, 1928). This theorem demonstrated that every finite two-person constant-sum game has a solution in pure or mixed strategies that guarantees each player at least a certain expected value.

In the case of mixed strategies, a player selects randomly from amongst its pure strategies, according to a particular probability distribution, which ensures itself of an expected value – called simply the *value* – regardless of what the other player does, as I illustrated in section 2.3 for the cardinal equivalent of game 44. (In this game, each player could ensure itself of a value of 2.5; in a zero-sum game, this value is positive for one player and negative for the other.) By choosing mixed strategies, a player can make itself inscrutable – to itself as well as to its opponent (at least before it makes its random choice) – and hence incapable of being exploited by an opponent, who might otherwise be able to discern some pattern in its play.

Unaware of the Minimax Theorem, Morgenstern saw the Holmes–Moriarty story as an illustration of a paradox in which "an endless chain of reciprocally conjectural reactions and counter-reactions . . . can never be broken by an act of knowledge but always only through an arbitrary act – a resolution" (Morgenstern, 1935, p. 174). While prescient in

moves, beginning at (1,3), by expecting a new confrontation but not foreseeing – for reasons explained in the text – that a disaster would befall him.

13 The subsequent analysis of literary works as constant-sum games is adapted from Brams (1993a) with permission.

recognizing the arbitrariness of the resolution, Morgenstern did not yet know its mixed-strategy form, even though mixed strategies had actually been calculated for specific games before the Minimax Theorem was proved (Dimand and Dimand, 1992).

Conan Doyle's resolution, on the other hand, was to make Holmes one whit more clever than Moriarty, ignoring that Moriarty himself might have been able to make an anticipatory calculation similar to Holmes'. Moreover, the matter does not end there: Holmes could have anticipated Moriarty; Moriarty, Holmes; and so on, leading to Morgenstern's "endless chain" of reasoning.

In his short story, "The Purloined Letter," Edgar Allan Poe carried Conan Doyle's resolution one step further by assuming that an extremely clever boy could always calculate exactly how far ahead less clever opponents would reason. Then, in a game in which this boy guessed whether an opponent was concealing an odd or an even number of marbles in his hand, the clever guesser would be able to anticipate his opponent, whether the opponent was a "simpleton" or someone of great cunning (but not greater than his own). Here is how the clever boy, according to Poe, was able to do this:

> When I wish to find out how wise, or how stupid, or how good, or how wicked is any one, or what are his thoughts at the moment, I fashion the expression of my face, as accurately as possible, in accordance with the expression of his, and then wait to see what thoughts or sentiments arise in my mind or heart, as if to match or correspond with the expression. (quoted in Davis, 1970, pp. 26–7)

Labeling this reasoning "tortuous," Davis (1990) points out that "the adversary can undo all the boy's labor by simply randomizing, in which case it will take nothing short of the Delphic Oracle to gain an edge." Davis (1990) used this example in his book (Davis, 1970) "because of the irony of Poe's comment: 'As poet and mathematician, he would reason well; as mere mathematician, he could not have reasoned at all'" (quoted in Davis, 1970, p. 27). On the contrary, Davis (1990) argues, that "as mathematician (using the minimax theorem) *he need not reason at all* – random play is sufficient to confound the boy."

Hence, it is the "mathematician" – who, according to Poe, "could not have reasoned at all" – who can play this game at least to a draw, even against an incredibly clever opponent. By randomizing, the mathematician robs the opponent of any control over the outcome and so ensures the value of the game.[14]

[14] Of course, knowing exactly how clever an opponent is, the boy can always win, but this cleverness is better characterized as "omniscience," which even the biblical God did not possess (Brams, 1980, 1983). In section 6.7 I discuss a paradox of omniscience.

This is a fundamental insight of the Minimax Theorem that neither Conan Doyle nor Poe seems to have understood. (To be sure, the cunning these writers attributed to their characters may make for better fiction than resolving each game with the flip of a coin.) But just the opponent's knowledge of this greater cunning would have been sufficient for them to even the score by choosing mixed strategies, according to the Minimax Theorem. Apparently they did not have even this knowledge – or, more accurately, the writers did not choose to give it to them.

Not all writers portray their characters in such a one-sided fashion. For example, knowledge is more shared, and calculations more even-handed, in the climactic scene of William Faulkner's novel, *Light in August* (published in 1932), in which Percy Grimm pursues Joe Christmas, a prisoner who has just escaped his captors. Though handcuffed in front, Christmas, like Grimm, has a gun. Grimm thinks, as the pursuit by bicycle and on foot nears its end, like a game theorist:

> He can do two things. He can try for the ditch again, or he can dodge around the house until one of us gets a shot. And the ditch is on his side of the house. (Faulkner, 1950, p. 404)

Grimm runs for the ditch, but soon he realizes that

> he had lost a point. That Christmas had been watching his legs all the time beneath the house. He said, "Good man." (Faulkner, 1950, p. 405)

The pursuit continues until it reaches the house of Reverend Hightower, who, though knocked down and injured by Christmas when Christmas burst in, refuses to tell Grimm in which room Christmas has hidden, despite Grimm's demands: "'Which room?' Grimm said, shaking him. 'Which room, old man?'" (Faulkner, 1950, p. 406). After Grimm asks once again, Hightower attempts to exonerate Christmas for the alleged murder he committed, but Grimm "flung the old man aside and ran on" (at random?) into the kitchen (Faulkner, 1950, p. 406).

A fictitious "Player" – a literary device in the novel that Faulkner inserts as a kind of supernatural character – guides Grimm, but this guidance seems founded on no more than chance moves. As Grimm storms into the kitchen, where Christmas has overturned a table to protect himself, he unloads his revolver into the table. Before Christmas dies, Grimm castrates him with the butcher knife he finds in the kitchen.

This, the most gruesome scene in the novel, contrasts sharply with Grimm's pursuit of Christmas, which is all cool calculation. Faulkner seems to have invented Player to epitomize the calm and deliberate mind of the fanatic; Grimm, who is "moved" by Player, is utterly devoid of any emotion, except when he explodes with savagery in the end. The beast in Grimm coexists with the cerebral Player, which is a juxtaposition that game theory normally does not entertain when it posits a player with one set of preferences.[15]

Unlike Conan Doyle and Poe, Faulkner beautifully captures the uncertainty inherent in mixed strategies, as well as how to act in the face of this uncertainty. And act Grimm does: first to his own disadvantage when he discovers that Christmas could follow his movements as he ran toward the ditch; second to his advantage when, "waiting for Player to move him again" (Faulkner, 1950, p. 406) – presumably in some randomized fashion – he rushes the kitchen. Faulkner has little to say about the motivations behind Christmas' choices, but it seems they were essentially arbitrary.

Faulkner does not assume that one player has superior computational abilities. True, Grimm has Player on his side, so to speak, but this device, in my view, reinforces the desultory character of Grimm's choices. Calculated they may have been, but because Grimm, at each stage of the pursuit, had only imperfect information, he could never be sure what his best choice was.[16] Grimm "won," finally, not because of sheer cleverness but because the game was unfair – the odds were heavily stacked against the fugitive, Christmas, whom Grimm so relentlessly hunted down.

I have offered this analysis of a scene from *Light in August* to suggest that Faulkner is one fiction writer who had an astute if implicit understanding of mixed strategies in two-person constant-sum games of imperfect information. Doubtless, other examples could be found. While the scenes that Morgenstern and Davis discussed in *Sherlock Holmes* and "The Purloined Letter" have the earmarks of games in which mixed strategies are optimal, both Conan Doyle and Poe shrank from making their protagonists' opponents as smart as the protagonists themselves. They got tidy results that way, but the minimax solution in games of imperfect information shows that not all conflicts can be resolved by outguessing. Faulkner understood this.

Faulkner also implicitly understood the implications of TOM. Christmas can go for the ditch or dodge around the house, and Grimm can rush the kitchen or some other room. If their total-conflict game is 44 (figure 2.3), which is the only 2×2 total-conflict game whose cardinal-equivalent

[15] If more than one type of player is allowed, as in games of incomplete information, only one type is actually the true type – there are not different types embodied in a single player (e.g., multiple personalities).

[16] Imperfect information is distinguished from incomplete information in games. When a player does not know at what point in the game play is at (e.g., what the players' previous choices were), its information is said to be *imperfect*. When a player does not know the rules or the payoffs of the players, its information is *incomplete*. Precise definitions of these terms can be found in the recent texts of Rasmusen (1989), Myerson (1991), Fudenberg and Tirole (1991), and Binmore (1992).

solution is in mixed strategies, both it and its AG offer no resolution unless the players know where they are in the game. But Grimm does not know, when he must choose, what choices Christmas had made, either early in the pursuit (ditch or house) or later (kitchen or other room), which is why his own response is necessarily arbitrary.

Both TOM and the standard theory recognize this anticipation problem, but TOM's framework, because it is ordinal, provides no basis for calculating mixed strategies. Perhaps serious poker players compute how often they will bluff, on average,[17] but I doubt that many players in real-life games make such a calculation and behave in a deliberately capricious manner.[18]

On the other hand, the standard theory does not distinguish between the three cyclical games with one NME (29, 30, and 31 in the Appendix) and the six cyclical games with two NMEs (42, 43, 44, 45, 46, and 47 in the Appendix), two of which I have used as models in this chapter (44 of elections, and 42 of the conflict between Pharaoh and Moses/God). It is the latter six games, in particular, in which anticipation is a problem – but only if the players do not know the initial state.

If they do, which I believe is often true, there is a resolution: the players will end up at one or the other of the two NMEs. But if the players do not start in a state but instead make simultaneous strategy choices, there is no resolution as to which will be chosen. At least, however, TOM says that the outcome can be narrowed down to these two.

There is an NME associated with every state in a 2×2 game, which I showed explicitly in figures 2.3 and 2.4 by juxtaposing a game and its AG.

[17] For game-theoretic models and advice on poker, see Ankeny (1981) and Binmore (1992, pp. 573–602).

[18] A strict ordinal game, whose cardinal solution is in mixed strategies, has been used to model the decisions that military commanders allegedly took in deploying their forces in the World War II Battle of Avranches; the model is described in Haywood (1954) and summarized in Brams (1975, pp. 13–24). This battle occurred just after the Normandy invasion in August 1944, but Ravid (1990) has shown that U.S. General Omar Bradley had access to ultra communications that revealed German plans to attack, making his decision to deploy his reserve of four divisions against the attack – knowing in advance German General von Kluge's choice – effectively a best response in a one-person game. Haywood (1954) also analyzed another World War II battle, the Battle of the Bismarck Sea between Allied and Japanese naval and air forces in February 1943, as a constant-sum game, which is also summarized in Brams (1975, pp. 3–10). But this game, which resembles game 25 in figure 2.2, has a saddlepoint according to the standard theory, and one NME according to TOM. If there is a good real-life example of the use of mixed strategies, I am not aware of it. Even Haywood (1954) eschewed the calculation of optimal mixed strategies in his Bismarck Sea example, presenting his game only as an ordinal one. Other approaches to expunging game theory of strategic uncertainty include Harsanyi (1977), Harsanyi and Selten (1988), and Güth and Kalkofen (1989), all of which assume a cardinal framework and propose different postulates for obtaining unique solutions to games.

But as I illustrated with game 56 in figure 2.3, a state may go into two NMEs and, therefore, be indeterminate.

This is a different kind of ambiguity than that to which cyclical preferences give rise in games with two NMEs. In section 5.1 I will show how the former kind of ambiguity – indeterminate states – can be resolved if one player possesses "order power." But next I shall explore why players may find it in their interest deliberately to forego their best payoffs and move, at the start, to a state where they receive less.

3 Magnanimity: it sometimes pays

3.1 Introduction

The anticipation problem I discussed in chapter 2 crops up only when players must choose strategies simultaneously. When players make such choices in the six 2×2 strict ordinal games vulnerable to this problem, it is impossible to determine which of the two NMEs will be selected in the AGs of these games. But this problem disappears when play commences in a state of the original game, which always goes into a single NME (though it may be indeterminate, which is a problem I take up in chapter 4).

In seven other 2×2 games, a different problem of anticipation occurs, even when play commences in a state. If the initial state is either (4,1) or (4,2), it may be rational for *R* to move – even though it obtains its best payoff by staying in this state, assuming that *C* also stays.

The problem is that it is rational for *C* to move in these seven games if *R* does not. But if *C* moves first, both players are worse off than had *R* been "magnanimous" – that is, moved initially to a state where it receives 3 instead of 4.

Game 28 in figure 3.1 illustrates this problem for *R*. Standard theory

		Column (*C*)		
		t_1	t_2	
Row (*R*)	s_1	(2,2) [3,4]	(4,1) [3,4]	← Dominant strategy
	s_2	(1,3) [3,4]	(3,4) [3,4]	

Key: (x,y) = (payoff to *R*, payoff to *C*)
$[x,y]$ = [payoff to *R*, payoff to *C*] in anticipation game (AG)
4 = best; 3 = next best; 2 = next worst; 1 = worst
Nash equilibrium in original game underscored
NME circled

Figure 3.1 Game 28

predicts the selection of its unique Pareto-inferior Nash equilibrium of (2,2), which is the product of *R*'s dominant strategy of s_1 and *C*'s best response – given *R*'s dominant choice – of t_1. TOM, on the other hand, predicts the selection of the NME of (3,4), which is Pareto-superior, wherever play starts.

I forego a complete derivation of the latter result, based on backward induction, and instead focus on what happens when (4,1) is the initial state:

	R		*C*		*R*		*C*		
R starts:	(4,1)	→|	(3,4)	→|	(1,3)	→	(2,2)	→	(4,1)
Survivor:	(4,1)		(3,4)		(2,2)		(2,2)		
	C		*R*		*C*		*R*		
C starts:	(4,1)	→	(2,2)	→|	(1,3)	→|	(3,4)	→	(4,1)
Survivor:	(2,2)		(2,2)		(1,3)		(4,1)		

In doing this kind of analysis earlier (sections 1.4 and 2.3), I assumed that *C*'s rational choice of moving to (2,2) takes precedence over *R*'s rational choice of staying at (4,1), making (2,2) the outcome.

But the relevant portion of rule 6 (section 1.4) covering this case is actually ambiguous on the matter:

> If it is rational for one player to move and the other player not to move from the initial state, then the player who moves takes *precedence*: its move overrides the player who stays, so the outcome will be that induced by the player who moves.

Is it rational, however, for *R* to move and induce (2,2) as the outcome? And is it rational for *C* to stay, anticipating this outcome?

The answer for both players to these questions is unequivocally no. To be sure, these choices are rational on the basis of backward induction by each player alone, but this one-sided analysis is flawed in game 28. It does not take into account *R*'s possibility of moving past its initial blockage at (4,1) to (3,4). If it does, there is again blockage at (3,4), but now it is in both players' interest that the process stop, compared to stopping at (2,2) when *C* moves first from (4,1).

In fact, *C*'s moving first to (2,2) is inconsistent with a portion of rule 5 (section 1.4):

> A player will not move from an initial state if this move (i) leads to a less preferred final state (i.e., outcome).

Clearly, (2,2) is less preferred than (3,4) not only by *C* but also by *R*.

But rules 5 and 6 can be reconciled with what I call the

> **Two-sidedness convention (TSC):** If one player (say, *C*), by moving, can induce a better state for itself than by staying – but *R*

by moving can induce a state Pareto-superior to C's induced state – then R will move, even if it otherwise would prefer to stay, to effect a better outcome.

Game 28 illustrates this convention. *R* would prefer to stay at (4,1), but recognizing that if *C* moves the outcome will be (2,2), it is in both *R*'s and *C*'s interest that *R* move and induce the Pareto-superior outcome of (3,4).

In section 1.4 I called rule 6 the "two-sidedness rule." I could make TSC rule 7, but I think it better to call it a "convention" because it clarifies the interpretation of the above-cited portions of rules 5 and 6. That is, it says under what conditions "it is rational for one player to move" (rule 6), but not on the basis of one-sided backward induction alone. Rather, it may be rational to move in order to deter the other player from making a prior move that "(i) leads to a less preferred final state (i.e., outcome)" (rule 5).

Implicitly, rule 5 (about moving only if one can improve the outcome) and rule 6 (about the precedence of moving over staying) – together with TSC, which clarifies when players will move to effect a better outcome – define an NME. That is, an NME is the product of the two-sided rationality calculations, which these rules and convention spell out, along with the usual rules of play (rules 1–4).

In section 3.2, I use game 28 to model the conflict between a mugger and a victim, showing that the victim should submit – even though the victim would prefer to resist if he or she can frighten away the mugger – and compare this outcome with data on actual choices by muggers and victims in robberies. In section 3.3 I begin a general analysis of magnanimity, which I apply to the aftermaths of wars. Specifically, I ask: When is it rational for the victor to be magnanimous? Here the issue is less for a player to give in (e.g., to a mugger) and more for it to give up some spoils (or other perquisites) of victory after a war in order to induce cooperation on the part of the vanquished.

In the game-theoretic analysis of magnanimity in section 3.4, I posit a "generic game," which subsumes 12 specific 2×2 strict ordinal games. Each of the four possible outcomes in the generic game, it turns out, may be a rational outcome in at least one of the specific games. Generalizing across the specific games, I derive several propositions about the rationality of magnanimity (and nonmagnanimity) in the generic game.

In section 3.5 I give a number of empirical examples of the aftermaths of nineteenth- and twentieth-century wars to illustrate the occurrence of different outcomes in the generic game. In section 3.6 I summarize the results and comment on sacrifices that a player may be driven to make not by the action of an opponent but by a recognition that, if it is not generous, the opponent will be motivated to take harmful action that will hurt both players.

3.2 Self-restraint by a victim in muggings

Mugging is an ever-present danger, especially of city dwellers.[1] As I use the term, a "mugging" is a sudden attack, with theft as the aim, which takes place on a street or in some other public place, by a lone perpetrator against a lone victim. Like all robberies, force or violence either is used or threatened in stealing money or other property. I assume that the victim is unarmed, but the mugger may be armed with a gun, a knife, or not armed at all. Statistics relating the use or nonuse of different weapons to the behavior of victims will be reported later.

A victim, I assume, may cry out for help, try to run away, fight the mugger, or quietly hand over money or other valuables. To simplify the analysis, I categorize the first three options of the victim as "resist" (R) and the last as "don't resist" ($\bar{R}$). The strategies of the mugger are "use force" (F) or "don't use force" ($\bar{F}$), as depicted in figure 3.2.

Before ranking the four possible states for each player, let me specify the goals of each. I assume that the victim's primary goal is to avoid injury, whereas its secondary goal is to keep its money and valuables. Its tertiary goal is to facilitate the capture of the mugger, the likelihood of which will be increased if the victim resists and, by doing so, attracts public attention.

The mugger's primary goal is to obtain money or other valuables from the victim, whereas its secondary goal is to avoid attention from passersby. Its tertiary goal is not to use force, so that if caught it will not face charges of assault as well as robbery.

These goals imply the following rankings of the four states, which I describe starting with the upper-left hand state in figure 3.2 and moving in a clockwise direction:

I **Fight:** (2,2). The victim resists, and the mugger uses force. The mugger gets the victim's money but may attract attention, whereas the victim is injured and loses its money. Nonetheless, the victim has achieved its tertiary goal of increasing the probability of the mugger's arrest.

II **Mugger fails:** (4,1). The victim resists the mugger, who is frightened away, and achieves all its goals. The mugger achieves only its tertiary goal.

[1] This section is adapted from a student paper of a former NYU undergraduate, John M. Parker.

		Mugger	
		Use force (F)	Don't use force ($\bar{F}$)
Victim	Resist (R)	I Fight (2,2) [3,4]	II Mugger fails (4,1) [3,4]
	Don't resist ($\bar{R}$)	IV Involuntary submission (1,3) [3,4]	III Voluntary submission (3,4) [3,4]

Key: (x,y) = (payoff to victim, payoff to mugger)
$[x,y]$ = [payoff to victim, payoff to mugger] in anticipation game (AG)
4 = best; 3 = next best; 2 = next worst; 1 = worst
Nash equilibrium in original game underscored
NME circled

Figure 3.2 Mugging game (game 28)

III **Voluntary submission:** (3,4). The victim gives up its money, and the mugger leaves the victim unharmed. The mugger achieves all its goals, whereas the victim achieves its primary goal of escaping unharmed.

IV **Involuntary submission:** (1,3). The victim gives up its money, but the mugger uses force anyway. The victim achieves none of its goals, whereas the mugger achieves its two most important goals, sacrificing only its tertiary goal by taking a greater risk of getting caught.

The resulting game is in fact 28, which I previously analyzed in section 3.1 (figure 3.1). According to the standard theory, the outcome will be (2,2), the Pareto-inferior Nash equilibrium.

In applying TOM, I assume that a move by the mugger from F *to* $\bar{F}$ is infeasible (section 1.6): once the mugger has used force, it cannot relent and not use it. If the game starts without the use of force by the mugger ($\bar{F}$) – at which point the victim may choose R or $\bar{R}$ – this infeasibility has no effect on the possible moves of the players. Then the initial states are (4,1) ("mugger fails") or (3,4) ("voluntary submission"). If the players start at (4,1), it is rational for the victim to move to (3,4), lest the mugger move to (2,2); if the players start at (3,4), it is rational for them to stay in this state. TOM also predicts (3,4) as the outcome when play commences at (2,2) or (1,3), but because moves that would bring the players to (3,4) from these states are infeasible, I do not make these predictions in the mugging game.

How does the prediction of (3,4), at least when the mugger starts out with $\bar{F}$, agree with statistics on muggings? Because "mugging" is not reported as a crime, I shall use statistics on the more general category of

*1 Unarmed robber (*n*=509)*

		Robber: Use force (F)	Robber: Don't use force ($\bar{F}$)
Victim	Resist (R)	39	6
	Don't resist ($\bar{R}$)	343	121

*2 Knife robber (*n*=226)*

		Robber: Use force (F)	Robber: Don't use force ($\bar{F}$)
Victim	Resist (R)	21	12
	Don't resist ($\bar{R}$)	70	123

*3 Gun robber (*n*=358)*

		Robber: Use force (F)	Robber: Don't use force ($\bar{F}$)
Victim	Resist (R)	17	24
	Don't resist ($\bar{R}$)	45	272

Source: Conklin (1972, p. 118, table 15)

Figure 3.3 Statistics on robberies in Boston, 1964 and 1968

robberies – probably most of which can be classified as muggings – in Boston in 1964 and 1968 (Conklin, 1972).[2]

The statistics in figure 3.3 are derived from Conklin (1972, p. 118, table 15) and give the numbers of robberies that correspond to the states shown in figure 3.2 for an unarmed robber, a robber with a knife, and a robber with a gun or other firearm.[3] Overwhelmingly, the victim chooses not to resist, with no resistance most frequent in the case of the unarmed mugger (91.2 percent), next in the case of the gun robber (88.5 percent), and least in the case of the knife robber (86.4 percent).

[2] I have proposed a model of muggings rather than robberies because the goals and preferences of the players can more easily be specified. Thus, in a bank robbery, a victim might prefer to resist in the face of force rather than submit in the face of no force if resistance involves only activating a silent alarm.

[3] There are also data on muggings in Great Britain (Pratt, 1980), but not in the form presented here. More recent data on robberies in the United States (Whitaker, 1989, p. 3, table 3) show that 27 percent of robbery victims attack the robber and 43 percent offer some kind of resistance, but these numbers are not broken down in terms of the kinds of weapons that the robber uses, as in Conklin (1972).

The great success of the unarmed robber in inducing compliance may seem paradoxical. It is explained by the fact that the vast majority of submissions in the unarmed case are because the robber uses force at the outset to intimidate or overcome the victim (Conklin, 1972, p. 116). Hence, submission is "involuntary." In fact, the unarmed robber uses force far more often (75.0 percent) – presumably at the outset[4] – than the knife robber (40.3 percent) or the gun robber (17.3 percent).

With the threat of violence more evident to victims in the case of knife or gun robberies than of unarmed robberies, it is not surprising that the two players choose "voluntary submission" more often than any other state. Indeed, in gun robberies this NME state is selected 76.0 percent of the time, compared with the selection of the Nash-equilibrium state of "fight" 4.7 percent of the time.

In the case of unarmed robberies, the NME of "voluntary submission" ranks only second (23.8 percent), behind "involuntary submission" (67.4 percent). But unarmed robberies, in my opinion, do not provide a proper test of the model, because only the victim is making a strategy choice if the robber chooses F at the outset. Since the robber cannot retract on the use of force, he or she is not a genuine player in a game, rendering the application of TOM dubious in this case.

Knife and gun robberies, by contrast, support the TOM prediction of "voluntary submission" in game 28. More important, TOM offers a rational interpretation of this choice that the standard theory does not. Although the data are silent on how the NME of "voluntary submission" is achieved – whether by F or $\bar{F}$ in the beginning – probably most robberies with weapons start and end in this state. However, some surely originate at (4,1) and then move to (3,4), once the victim realizes that continued resistance will lead to the Pareto-inferior state of (2,2).

The self-restraint of the mugging or robbery victim, of course, is a very different matter from the magnanimity of the victor in a war. Nevertheless, each involves some sacrifice on the part of a player to try to head off a still worse outcome.

3.3 Different views on the rationality of magnanimity

A dilemma facing every victor in an interstate war is how to treat the vanquished opponent when hostilities end.[5] Should the victor strive for a postwar settlement that addresses at least some of the grievances of the

[4] When F is chosen by the robber initially, I previously assumed that a switch to $\bar{F}$ is infeasible. However, the data do not distinguish the robber's initial choice of F from its initial choice of $\bar{F}$, whence it switches to F in the course of the robbery.

[5] This and the next three sections are adapted from Brams and Mor (1993) with permission.

vanquished, or should it implement a new status quo that does not acknowledge these grievances? Magnanimity may quell the desire of the vanquished for revenge, but nonmagnanimity may prevent the vanquished from acquiring the means to mount future challenges.

If, as Clausewitz (1832) argued, wars are fought over the preferred political order, the vanquished's position in this order, after the war, cannot be ignored. Several scholars of international relations have analyzed the vanquished's role in the postwar stability of a system. Kissinger (1964), for example, concluded from his analysis of early nineteenth-century Europe that restoring stability to a postwar system requires magnanimity toward the vanquished opponent by the defenders of the status quo. Oren (1982, p. 150), focusing on two-state conflicts, also assembled evidence that "prudence in victory" is more stabilizing than punitive behavior by the victor, because the latter strategy produces a desire for revenge on the part of the vanquished unless it is annihilated.

Aron (1966), by contrast, advanced a "peace by empire" argument, claiming that postwar stability is served better by a total subjugation of the opponent, which robs it of the means, and hence the opportunity, to initiate future conflicts. Maoz (1984), using aggregate data on serious interstate disputes, tested the contradictory Oren and Aron theses and found empirical support for the latter. Later Maoz (1990, pp. 256–7) offered a rationale for "peace by empire" by showing this to be the Nash equilibrium in a specific 2×2 nonstrict ordinal game (i.e., with some preferences of one player tied).

In the remainder of this chapter, I approach the problem of magnanimity from a different perspective – namely, that the victor's dilemma cannot be resolved exclusively in favor of magnanimity or nonmagnanimity. I show the conditions under which "prudence in victory" on the one hand, and "peace by empire" on the other, are rational, given possible counteractions that the defeated party may take.

This perspective assumes that the defeated party is not inert but itself a player in a game. (True, there have been wars in which the defeated party was utterly devastated and, therefore, incapable of making any choices, but these are relatively rare.) The postwar political order is thereby the product of *interdependent* choices of the victor and the defeated party, with the players' preferences a function of both the victor's choice of magnanimity or nonmagnanimity and the defeated party's choice of cooperation or noncooperation. To analyze the rationality of these strategic choices, I define a generic "Magnanimity Game," which subsumes different strategic situations that may arise in the aftermath of victory.

These situations reflect the different preferences that the victor and the defeated party may have for the four possible outcomes in this game.

Thus, rather than positing a specific game between the victor and the defeated party, I allow for two types of victor and six types of defeated, which defines 12 specific games. I then show when magnanimity or non-magnanimity by the victor, and cooperation or noncooperation by the defeated party, are NMEs in each of the games and draw some generalizations across the games.

I assume that the postwar situation, in the wake of victory, defines the initial state. The NMEs that may arise from this state split into three classes of games, which together allow for each possible state to be the outcome.

I then apply the Magnanimity Game to several historical cases – mostly the aftermaths of wars in the nineteenth and twentieth centuries – to illustrate, empirically, the different possible outcomes in the game. I conclude by discussing some larger implications, including normative, of the analysis.

3.4 The Magnanimity Game (MG)

Consider the aftermath of a war or other major international dispute, such as a crisis, in which one player, the victor (V), prevails over another player, the defeated (D). In the postdispute situation, assume V has a choice of being either magnanimous (M) or not magnanimous ($\bar{M}$) to D, and D has a choice of either cooperating (C) or not cooperating ($\bar{C}$) with V.

In section 3.5 I shall discuss the meaning of these choices, and the resulting outcomes, in some historical cases. For now assume that, immediately after the dispute, the players are at Status Quo in the payoff matrix of figure 3.4. In this state, V is in its best position and D is in an inferior position – that is, there is at least one other state that D would prefer.

To give further structure to this postdispute situation, I make some additional assumptions about how the players rank the various states in figure 3.4. In this representation, the higher the numerical subscripts of v and d, the greater the payoffs. How these payoffs compare with the payoffs having lettered subscripts is indicated below, moving counterclockwise from the upper-left hand state in figure 3.4:

I **Status Quo.** Best for V (v_4) and inferior to Magnanimity for D ($d_i < d_{i+}$).

II **Magnanimity.** Next-best for V (v_3) and superior to Status Quo for D ($d_{i+} > d_i$).

III **Rejected Magnanimity.** Inferior for V (v_t, where $t = 1$ or 2, i.e., this outcome is either worst or next worst) and superior to Rejected Status Quo for D ($d_{j+} > d_j$).

		Defeated (D)	
		Cooperate (C)	Don't cooperate ($\bar{C}$)
Victor (V)	Don't be magnanimous ($\bar{M}$)	I Status Quo (v_4,d_i)	IV Rejected Status Quo (v_s,d_j)
	Be magnanimous (M)	II Magnanimity (v_3,d_{i+})	III Rejected Magnanimity (v_t,d_{j+})

Key: $(x,y)=$(payoff to V, payoff to D)
$v_4 > v_3 > v_s,\ v_t\ (s,\ t=1 \text{ or } 2)$
$d_{i+} > d_i;\ d_{j+} > d_j$

Figure 3.4 Magnanimity Game (MG)

IV **Rejected Status Quo.** Inferior for V (v_s, where $s=1$ or 2, i.e., this state is either worst or next worst) and inferior to Rejected Magnanimity for D ($d_j > d_{j+}$).

These rankings, because they do not give a complete ordering of states from best to worst for each player but rather only a partial ordering, do not define a specific ordinal game but rather a *generic game*. This game, which I call the *Magnanimity Game* (MG), subsumes the preference orderings of 12 specific games (to be given shortly). It is meant to characterize postdispute situations generally by allowing for all reasonable possibilities of preference orderings by the two players.

Thus, when D chooses $\bar{C}$ I leave unspecified whether V prefers Rejected Magnanimity (v_t) to Rejected Status Quo (v_s). V might prefer the former if its generosity cannot be seriously exploited by an "ungrateful" D and will ultimately redound to V's favor. By contrast, if V's generosity creates an opportunity for D to recoup its losses and fight V again, V might prefer to clamp down by being nonmagnanimous.

In either event – whether V is magnanimous or not – I assume that V is always better off when D cooperates ($v_4 > v_3 > v_s$, v_t, where $s, t=1$ or 2). Not only is V always better off, but I also assume that Magnanimity (v_3) is next best to Status Quo (v_4). Thus, V gives up something to improve D's lot in moving from Status Quo to Magnanimity, but V suffers still more (v_1 or v_2) when D chooses $\bar{C}$.

The partial ordering of payoffs I assume for D is less complete than that assumed for V. Because D prefers its payoffs with "plus" subscripts to comparable payoffs without the plus, D desires that V always be magnanimous. But because I assume no ordering between the i and $i+$ payoffs on the one hand, and the j and j+ payoffs on the other, I cannot make any comparisons between D's payoffs associated with C versus $\bar{C}$.

1 Status Quo is NME (4 games)

22a		33		34		35	
(4,2)	(2,1)	(4,3) [4,3]	(2,1) [4,3]	(4,3) [4,3]	(1,1) [4,3]	(4,2) [4,2]	(2,1) [4,2]
(3,3)	(1,4)	(3,4) [3,4]	(1,2) [4,3]	(3,4) [3,4]	(2,2) [4,3]	(3,4) [3,4]	(1,3) [4,2]

2 Magnanimity is NME (4 games)

28		29		Prisoners' Dilemma 32		50	
(4,1)	(2,2)	(4,1)	(1,2)	(4,1) [3,3]	(2,2) [2,2]	(4,2) [3,4]	(1,1) [3,4]/[4,2]
(3,4)	(1,3)	(3,4)	(2,3)	(3,3) [3,3]	(1,4) [3,3]	(3,4) [3,4]	(2,3) [4,2]/[3,4]

3 Rejected Status Quo or Rejected Magnanimity is NME (4 games)

Total conflict 11		18		22b		Chicken 57	
(4,1)	(2,3)	(4,1)	(1,3)	(4,1)	(1,2)	(4,2) [3,3]/[2,4]	(1,1) [2,4]/[4,2]
(3,2)	(1,4)	(3,2)	(2,4)	(3,3)	(2,4)	(3,3) [3,3]	(2,4) [4,2]/[3,3]

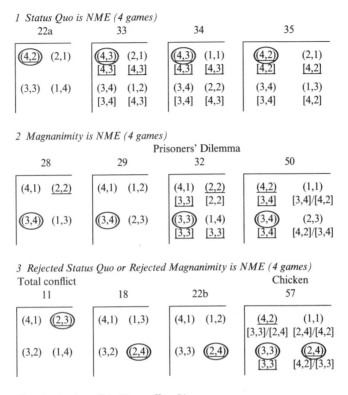

Key: (*x*,*y*) = (payoff to *V*, payoff to *D*)
[*x*,*y*] = [payoff to *V*, payoff to *D*] in anticipation game (AG)
4 = best; 3 = next best; 2 = next worst; 1 = worst
Nash equilibria in original games and AGs (only those with more than one NME shown) underscored
NMEs, when Status Quo is initial state, circled

Figure 3.5 12 specific ordinal games subsumed by MG

These partial orderings of the players in MG do not render any state a Nash equilibrium, much less endow either player with a dominant strategy, in the standard theory. Neither, according to TOM, is any state of MG an NME.

To determine what states may be NMEs, I show in figure 3.5 the 11 specific game configurations subsumed by MG. Actually, 12 games are shown in figure 3.5, because game 22 is shown twice (once in class 1 with an "a" after its number, and once in class 3 with a "b" after its number), with *V* and *C* as well as their strategies interchanged.

Even though a and b are the same game configuration (i.e., game 22), their players receive different payoffs in Status Quo (state I in figure 3.4). Thus in game 22a, *D* receives a payoff of 2 in this state, whereas *D* receives a payoff of 1 in game 22b.

This fact makes these games different in the present analysis, in which I assume that play always starts at Status Quo, just after the conclusion of a dispute. Each player must then decide whether to depart from (v_4, d_i) or not.[6]

I have circled the outcomes in figure 3.5 which are NMEs from Status Quo. If a game contains more than one NME, they are indicated in brackets below the initial states from which they originate in the anticipation game (AG) (section 2.3). Thus in game 32 (Prisoners' Dilemma), Status Quo of (4,1) goes into (3,3), which is circled, but (2,2) goes into itself, which makes it another NME but not one that is induced from Status Quo.[7]

In figure 3.5, I have divided the 12 specific games subsumed by MG into three classes:

I **Status Quo is NME (four games).** It is rational for *V* not to be magnanimous, because *D* will not depart from Status Quo.

II **Magnanimity is NME (four games).** It is rational for *V* to be magnanimous, because otherwise *D* will depart from Status Quo – but not from Magnanimity. These games include the infamous Prisoners' Dilemma (game 32), about which I shall say more in section 5.3.

III **Rejected Status Quo or Rejected Magnanimity is NME (four games).** Even being magnanimous does not prevent the choice of $\bar{C}$ by *D* in these games, which include Chicken (game 57), another infamous game that I shall discuss in section 5.3.[8] Rejected Magnanimity is rational in three of these games (18, 22b, and 57), whereas Rejected Status Quo is rational in the remaining game (11), which is the only game of total conflict among the 12 specific games subsumed by MG. Although there is an even 4–4–4 split of the specific games among the

6 Brams and Mor (1993) give necessary and sufficient conditions for the departure (or nondeparture) of each player from Status Quo to each of the other states. Their analysis is based on somewhat different rules than those given here, but the results are the same as those for TOM, except for game 57 (Chicken), in which their rules identify only (2,4), not also (3,3), as the NME from Status Quo.

7 With the exception of Chicken (game 57), all the NMEs induced from Status Quo are Nash equilibria in the AG, but there may be other Nash equilibria in the AG. Indeed, every entry in the AG of a one-NME game is a Nash equilibrium.

8 Because of its indeterminate NME of (3,3)/(2,4) at Status Quo, which includes (3,3), Chicken also falls into class 2; for convenience I list it only in class 3 because of its NME of (2,4). However, Chicken's NME of (3,3) is different from the other class 2 Magnanimity NMEs. In Chicken, *V*'s move to Magnanimity at (3,3) is rational because *D*, by moving first, can induce (2,4), which is better for itself and worse for *V*. By contrast, in the other class 2 games, *V*'s move to Magnanimity is rational because *D*, by switching to $\bar{C}$, can induce a Pareto-inferior outcome that hurts both players.

three classes (but see note 8), these numbers may bear little relation to the empirical frequency with which Status Quo, Magnanimity, or Rejected Status Quo and Rejected Magnanimity are actually chosen in real-life games.

I next offer some more general results, in the form of three propositions, that follow from the analysis of the 12 specific games:

> **Proposition 3.1.** *It is rational for V to be magnanimous in seven games (the NMEs in all class 2 games, and games 18, 22b, and 57 in class 3, are associated with this strategy) and not magnanimous in five games.*

It is worth noting that the NMEs from Status Quo in the 12 games do not always coincide with the Nash equilibria based on the standard theory. True, the NMEs coincide with Nash equilibria in class 1 and class 3 games [except for (3,3) in Chicken (game 57)], but in class 2 games, including Prisoners' Dilemma (game 32), the NME of Magnanimity is never a Nash equilibrium. In fact, Magnanimity is not a Nash equilibrium in any of the 12 specific 2×2 games, which makes this outcome inexplicable as a rational choice, according to the standard theory – unless, of course, MG fails to capture the postwar strategic situation.

Starting at Status Quo, *V* will move to Magnanimity in all class 2 games. In game 32 (Prisoners' Dilemma) and game 28 (discussed in sections 3.1 and 3.2), in particular, Magnanimity obviates the choice of the Pareto-inferior (2,2) Nash equilibrium in these games. More generally, one can readily verify

> **Proposition 3.2.** *The NME in all 12 games is Pareto-optimal – there is no other outcome better for both players.*

Finally, observe in figure 3.5 that all games with (4,3) "cooperative" Status Quos are in class 1, whereas no games in class 1 have (4,1) "total-conflict" Status Quos, giving

> **Proposition 3.3.** *Status Quo is never an NME when it is D's worst (1) state, but it is always an NME when it is D's next-best (3) state.*

In other words, the better off *D* is at Status Quo, the more likely this state is an NME. Status Quo may or may not be an NME when it is an "intermediate" (4,2) – it is the NME in two games (22a and 35) but not in two others (50 and 57).

3.5 Applications of MG to historical cases

My discussion of historical cases in this section is meant to be illustrative and not a systematic test of the predictions of MG. An empirical test would require ascertaining the specific preferences of players in the aftermaths of particular wars and trying to corroborate that they made strategy choices consistent with TOM.

I will give several examples in which the four different states of MG, all of which may be NMEs, appear to have been selected as outcomes. These examples, I believe, lend plausibility to MG as an aid in thinking about the conditions that give rise to magnanimity or nonmagnanimity on the part of the victor, and its acceptance or rejection by the defeated.

Status Quo. One war from the nineteenth century and two from the twentieth illustrate this outcome. After France's defeat in the Franco-Prussian War of 1870–1, Prussia humiliated France by issuing a proclamation of the new German empire at the Versailles Palace of Louis XIV, which it followed with a victory parade through the streets of Paris. More significant, France was required to pay an indemnity of five billion francs (with German occupation troops remaining until payment was completed, which occurred in 1873); and the French provinces of Alsace and Lorraine were annexed to Germany.[9]

While Chancellor Otto von Bismarck successfully opposed some of the more extreme demands his army generals made against France in 1871, it is clear that Germany chose $\bar{M}$ and France could do little but swallow its defeat. Indeed, contemplating their day of revenge, which was to come at Versailles almost fifty years later (1919), the French coined the rueful phrase, "Never speak of it, always think of it" (Ziegler, 1987, p. 17).

In the aftermath of World War I, not only was Germany required to surrender unconditionally, but it was also forced to accept the harsh terms of a settlement imposed by the allies. Likewise, the surrenders of Nazi Germany and Japan at the end of World War II were unconditional, with Germany this time divided into four zones. Once again, the allies made no concessions after the war, although the Marshall Plan, beginning in 1947, helped tremendously in the later reconstruction of Europe, including West Germany.[10]

There was no Marshall Plan after World War I. Germany, plagued by inflation and destitute, moved inexorably toward Nazism, especially after 1933. One might contrast the aftermaths of the two world wars by saying that, immediately after both, the allies chose $\bar{M}$ and Germany could do no better than choose C, giving Status Quo. But a subsequent shift toward M by the allies after World War II engendered West Germany's continuing choice of C (Magnanimity), whereas the lack of such a shift by the allies

[9] In its successful war against Austria five years earlier (1866), Prussia, by contrast, treated Austria much more leniently, as I discuss shortly.

[10] Of course, the aid provided was in part spurred by U.S.–Soviet rivalry after the war.

after World War I led Germany eventually to choose $\bar{C}$ (Rejected Status Quo).

In summary, the immediate aftermath of the Franco-Prussian War and both world wars was Status Quo. By the dawn of World War I in 1914, however, the 1871 Status Quo had probably deteriorated to Status Quo Rejected – to the detriment, ultimately, of both France and Germany, both of which suffered horrendous losses in World War I.

Likewise, the Status Quo of 1918 became Rejected Status Quo some twenty years later, with the world on the verge of World War II. By comparison, only ten years after World War II, the outcome had certainly become Magnanimity, with West Germany joining NATO in 1955 and Japan becoming part of the Western alliance, too. On the other hand, the main victors of World War II had parted ways to become rivalrous superpowers.

Magnanimity. The Magnanimity outcome after World War II became a reality only some years after the war, which raises the question of what time span the model supposes. This is an empirical question to which different answers are possible, depending on one's conception of how long a postdispute situation exists after a conflict.

I have no strong opinion about this, but there are certainly examples of the choice of Magnanimity immediately after a war. Consider the behavior of the Prussians after their victory in the Seven Weeks War (1866) against the Austrians, which I alluded to earlier (note 9):

> After defeating the Austrian army at the decisive battle of Sadowa, the Prussian army did not pursue the Austrian army across the Danube River. It did not hold a victory parade through the streets of Vienna ... [and it] did not annex portions of Austria near its own borders, even though some justification could have been made for incorporation. (Ziegler, 1987, p. 15)

What was the rationale of Bismarck's choice?

> This policy of restraint was achieved by some effort on Bismarck's part, against the desires of the king and some of his advisers. Bismarck realized, as the king did not, that the work of German unification was not yet completed, and a humiliated and bitter Austria would be a potential ally for the new obstacle that now stood in Prussia's way, France. (Ziegler, 1987, p. 15)

Prussia thereby headed off Austria's choice of $\bar{C}$ by instead moving to M, unlike its later choice of $\bar{M}$ against France. Indeed, the short duration of the Seven Weeks War can in part be explained by Bismarck's limited war aim, which was to prevent Austrian influence in Prussian affairs – and not acquire territory, as Prussia later did against France.

A more recent example of a magnanimous victor is David Ben Gurion after Israel's War of Independence. Although in late 1948 and early 1949 Israeli military capabilities permitted further territorial expansion to the north, east, and south, Ben Gurion "brought the war to an end and drew the lines delineating the new state under the guidance of the precepts of prudent moderation" (Oren, 1982, p. 155). Most noteworthy was his decision to withdraw Israeli troops from the Rafa heights, a key strategic point on the southern front, over the strong objections of his army staff (Sachar, 1979, p. 346).

The 1962 Cuban missile crisis is arguably another case ending in Magnanimity. Although the Soviet agreement to withdraw their missiles from Cuba was an outcome that President John Kennedy could present as an American victory, his brother, Robert Kennedy, wrote that

> after it was finished, he made no statement attempting to take credit for himself or for the Administration for what had occurred. He instructed all members of the Ex Comm and government that no interview would be given, no statement made, which would claim any kind of victory. (Kennedy, 1969, pp. 126–7)

And in his subsequent public statements on October 28 and November 2 and 28, President Kennedy refrained from claiming victory and emphasized instead the broad themes of peace in the Caribbean and the reduction of world tensions.[11]

Rejected Status Quo. The 1979 Soviet invasion of Afghanistan illustrates this outcome. After the Soviets captured Kabul, executed the Afghan president, and installed their own puppet, the Afghan rebels continued to resist, waging a long and costly guerrilla war – with considerable help from the United States – that eventually forced the Soviets to withdraw in 1988 and the Afghan government to collapse in 1992.

Similarly, although Israel managed to defeat and eject the Palestine Liberation Organization (PLO) from Lebanon after its 1982 invasion, raids by the PLO have continued to this day. On the other hand, Egypt and Israel chose Magnanimity after the 1973 Yom Kippur war, which I shall explain in terms of moving power in chapter 4.

Rejected Magnanimity. This outcome seems the most difficult to document, because magnanimity, like beauty, is in the eye of the beholder. Thus, if V claims it was magnanimous, D can respond that this was not so, and, hence, by not cooperating it never "rejected" magnanimity. None-

[11] On the other hand, Secretary of State Dean Rusk's statement, "We're eyeball to eyeball, and I think the other fellow just blinked" – spoken at the climactic moment of the crisis and reported several weeks later – does sound suspiciously like a victory claim.

theless, I present two cases that seem to illustrate the selection of this outcome.

After its invasion of Cyprus on July 20, 1974, Turkey could credibly claim victory on August 16, having gained control of 40 percent of the island. On February 13, 1975, Turkish Cypriot leaders declared a separate state on the northern part of the island. At the same time, they offered Greek Cypriots a confederacy, with power to be shared equally in a single state (though Greek Cypriots outnumbered Turkish Cypriots by four to one). This offer, whose "magnanimity" might be questioned, was rejected, and a formal settlement of this conflict has yet to be achieved.

A quick settlement was attempted by Saddam Hussein after his invasion of Iran on September 22, 1980. Although his forces encountered only feeble and disorganized resistance initially, Hussein did not capitalize on his advantage. Instead, he "voluntarily halted the advance of his troops within a week after the onset of hostilities, and then announced his willingness to negotiate an agreement" (Karsh, 1989, p. 211). Not only was this offer summarily rejected by Ayatollah Khomeini, but the war ground on for nearly eight more years.

Because the Iran–Iraq war had just commenced at the time of Hussein's offer, this case is not strictly comparable to earlier cases, in which a decisive military victory had been achieved. Indeed, Hussein's offer and Khomeini's rejection may be closer to the Rejected Status Quo cases of the Soviet invasion of Afghanistan and the Israeli invasion of Lebanon. Although the initial victories in the latter cases were more consequential, the aftermath of each invasion was continued fighting that blurred the identification of a victor and a defeated party.

3.6 When is sacrifice rational?

To summarize, I began by showing that in game 28 it is rational for *R* at (4,1) to sacrifice its best payoff in order to induce (3,4), lest *C* induce (2,2) – the unique Pareto-inferior Nash equilibrium – by moving first instead. I then used game 28 to model a game between a mugger and a victim and showed that robbery statistics generally confirm TOM's prediction of (3,4), the unique NME in this game.

A two-sidedness convention (TSC) clarified the application of rules 5 and 6 to give this result, which is relevant to a total of seven 2×2 strict ordinal games (27–31 with one NME, and 32 and 50 with two NMEs, listed in the Appendix). Four of these seven games are subsumed by the Magnanimity Game (MG) (all class 2 games in figure 3.5), which is a generic game that comprises a total of 12 specific 2×2 games.

To introduce MG, I began by discussing two contending schools of thought on how a victor should treat a defeated party after a war or other major dispute. Instead of taking sides in this controversy, I showed that each side may be "right," based on two-sided analysis in which the defeated party as well as the victor is considered a player able to make choices. Whether magnanimity or nonmagnanimity by the victor is rational depends on the specific game being played or, more generally, the class of games into which it falls.

The analysis of MG demonstrated that each of the four states of this generic game, starting at Status Quo, may be an NME. For example, sticking with the Status Quo is never rational when this state is the worst for the defeated party, but it is always rational when it is the defeated party's next-best state. Magnanimity is rational in some games, including Prisoners' Dilemma (game 32) and the mugging game (28), in which the Nash equilibria are Pareto-inferior. Noncooperation by the defeated party is a rational strategy in still other games, including a total-conflict game (11) and Chicken (game 57, in which cooperation may also be rational), with the resulting outcome either Rejected Status Quo or Rejected Magnanimity.

These findings may be viewed normatively, at least insofar as they offer interpretable conditions for deciding a rational course of action after a war or other major crisis. Thus, for example, because all the rational outcomes in MG are Pareto-optimal, the model may help decision makers avoid the Pareto-inferior Nash equilibria that occur in games like Prisoners' Dilemma and game 28. In addition, TOM may help decision makers look ahead and anticipate dire long-term consequences that may arise from seemingly attractive, yet short-sighted, actions.

To conclude, MG helps explicate, in a systematic fashion, the logic of strategic choices in postdispute situations. Because players' choices in these situations may critically affect the future stability of the international system, they need to be carefully thought through, which I believe TOM facilitates.

More generally, the analysis of this chapter demonstrates that self-restraint in a mugging, magnanimity after a war, or a cooperative gesture in other situations may, apart from reasons of altruism, make good strategic sense. Sacrifice may be rational not because it does not hurt the sacrificer but because it heads off action by the other player that hurts both players even more.

4 Moving power: breaking the cycle

4.1 Introduction

The nature of postwar settlements – and, in particular, whether the victor should be magnanimous or not – depends not only on the specific game being played but also on the rules of play and the rationality rules, which I discussed in chapter 1. In this chapter I will modify the latter rules by introducing an asymmetry in the capabilities of the players, which enables one player to continue moving when the other player must eventually stop.

This kind of power, which I call "moving power," was Egypt's strength in its protracted struggle with Israel from 1948 until 1978–9.[1] In each of the five wars that Egypt and Israel fought before the 1978 Camp David accords and their 1979 peace treaty (1948, 1956, 1967, 1969–70, and 1973), Israel prevailed but was unable to stabilize a postwar Status Quo, despite the plethora of military and diplomatic agreements, U.N. resolutions, and mediation efforts that followed each war (Touval, 1982). Instead, recurring cycles of violence and short-term accommodations characterized the prolonged conflict between these two antagonists. If this conflict were a class 2 game in figure 3.5, in which Magnanimity is the unique NME, it should not have taken 31 years to achieve peace.

This is a puzzle that I shall attempt to explain informally in this section and then more formally in section 4.3. However, I shall not try to encapsulate the 31-year conflict between Egypt and Israel in a single 2×2 strict ordinal game. Rather, I shall argue that Egypt's moving power, as the defeated party (D) in the Magnanimity Game (MG) (section 3.4), enabled it to induce Magnanimity as the outcome in a class 1 MG game. In the absence of moving power, Status Quo would be the outcome (i.e., the NME) in this class of games, given that play starts in the Status Quo state.

The cycles of Egyptian–Israeli violence, I believe, are explicable as a

[1] Part of this section is adapted from Brams and Mor (1993) with permission.

consequence of Israel's failure to acknowledge Egypt's moving power until after the shock of the 1973 Yom Kippur war. In MG, I interpret *D*'s moving power as giving it the ability to engage the victor (*V*) in repeated military confrontations, from minor skirmishes to major wars, even though *D* may chronically lose.

The cycling terminates when *V* finally realizes that *D* has the resolve and wherewithal to revisit the Status Quo again and again. Then it becomes rational for *V* to choose *M*, accepting Magnanimity as the outcome when *D* chooses *C*, rather than have *D* move to $\bar{C}$ and for the game, once again, to cycle back to Status Quo.

In fact, the repeated defeats suffered by the Egyptians did not deter them from rejecting each new settlement and confronting the Israelis on the battlefield again. That the Egyptians believed they possessed moving power is echoed in the comments of Hasannin Haykal, an Egyptian journalist and confidant of Egyptian president Gamal Abdel Nasser, who wrote in *al-Ahram* on March 16, 1962, that

> the more the independent strength of the UAR [United Arab Republic, which included Egypt and Syria] grows, the less will be the proportion that has to be devoted to meeting the Israeli danger. The opposite is the case on the other side of the barricade; the more the power of the UAR grows, the greater is the effort that Israel will have to make. (cited in Harkabi, 1972, p. 88)

Nasser, too, exuded confidence in Egyptian resilience when he said that "the present and the future do not work in her [Israel's] favor but in favor of the Arabs" (Harkabi, 1972, p. 88).

The Egyptian–Israeli series of wars were formally ended, six years after the 1973 Yom Kippur war, as a result of a mutual adjustment in the parties' perceptions of their relative power. Although victorious in yet another war in 1973, Israel came to realize that repeated defeats of Egypt could not guarantee an end to their conflict. Egypt, in effect, was indefatigable, politically if not militarily.

Why did this realization take so long? Over the years Israel's key leaders came to believe that only repeated demonstrations of Israeli military might would bring the Arabs to the negotiation table. For example, David Ben Gurion, Israel's first prime minister,

> concluded that there was no chance of reconciliation until Israel's strength and stability become so manifest that the Arab states would reconcile themselves to our permanence. In the meantime, he had not thought it wise to invest very deeply in contact throughout the Arab world. (Eban, 1977, p. 306)

Curiously, though, this view did not prevent Ben Gurion from being magnanimous after the display of Israeli military prowess, as I indicated in the case of the 1948 war (section 3.5). It seems that Israel's resounding victory in 1948 made it appear so strong that it did not feel it had to protect its reputation, as was the case later when its vulnerabilities – especially to terrorism – became apparent.

The human and economic cost imposed on Israel by the 1973 war – coupled with her isolation, her evident dependence on the United States, and the success of the Arab oil embargo – brought about a "rude awakening from a sweet but unreal dream" (Yaniv, 1987, p. 188). On the Egyptian side, Anwar Sadat's perception that Israel was ready to acknowledge Arab moving power and accept Magnanimity in MG allowed him to approach the Israelis on an equal footing – despite the defeat of his military forces in the Yom Kippur war, which were saved from disaster only by U.S. intervention. In the end, then, Egypt's moving power overrode Israel's military superiority, prompting a joint recognition by the players that only Magnanimity in MG could save them from themselves.[2]

In section 4.2 I postulate a new rationality rule (rule 5′ instead of rule 5), which permits players to cycle in a matrix, and then distinguish three classes of cyclic games. I postulate a second new rationality rule (rule 6′ instead of rule 6) in section 4.3, where I formally define "moving power" and show in which cyclic games it is "effective," "irrelevant," or "ineffective." If a game is not cyclic, moving power is not defined, which is true of the 12 symmetric 2×2 games, including Prisoners' Dilemma and Chicken, as well as 30 nonsymmetric games, or a total of 42 of the 78 2×2 games (54 percent).[3]

In section 4.4 I examine the effects of moving power in a "Revelation Game," played between an ordinary being and a superior being, and show how either player – but, more plausibly, the superior being – can implement its preferred outcome with moving power. Although certain moves would appear infeasible in this game, this is not the case if this game is played over a sufficiently long period of time so that the identity of the ordinary player changes (the superior being is assumed to be immortal).

In section 4.5 I examine the effects of moving power in the Vietnam war, especially as manifested in the repeated bombing campaigns that the United States undertook to try to wear down North Vietnam to the point where it would agree to a cessation of hostilities. North Vietnam, however, mounted ground offenses to try to convince the United States that it could hold out indefinitely, despite the bombing, in part because of

2 Of course, the players had no knowledge of MG, as defined in section 3.4, but my argument is that they implicitly understood its strategic implications, including the effects of moving power that I shall develop more fully in section 4.3.

3 A formal definition of a "symmetric games," in which players face the same strategic choices, is given in section 4.2.

an apparent misperception of U.S. preferences in the game they played at the end of the war.

In section 4.6, I conclude with some observations on how moving power makes itself felt. I argue that a player's reputation may obviate the need for it to make physical moves to demonstrate its tenacity or strength. But reputation alone did not suffice for either Israel in its conflict with Egypt, or the United States in its conflict with North Vietnam, in part because of incomplete information that led to misperceptions, a theme I shall return to later.

4.2 Cyclic games and rule 5′

As I noted in section 1.4, the rules of play (i.e., rules 1–4) say nothing about what causes a game to end, only when.[4] Rule 5, which forbids moves that do not lead to a better outcome and thus precludes cycling back to the initial state, provides one answer. But this ban on cycling may not be realistic, as the protracted Egyptian–Israeli conflict, in which the players did revisit the past again and again, makes unmistakably clear.

To try to capture the cyclic aspect of certain conflicts, I next define a class of games in which cycling is possible by precluding a class of games in which it is not. Rule 5′ provides a sufficient condition for cycling not to occur:

5′ If, at any state in the move–countermove process, a player whose turn it is to move next receives its best payoff (i.e., 4), it will not move from this state.

Rule 5′, in fact, precludes cycling in 42 of the 78 2×2 games, 21 of which contain a mutually best (4,4) state. As an illustration of a game that does not cycle, as well as one that does, consider the two class 1 MG games, 22a and 35, in figure 3.5, which are reproduced in figure 4.1. Starting from (4,2) in each of these games, which is Status Quo in MG, neither player has an incentive to move, according to TOM. Hence, (4,2) is an NME in each game.

But, in fact, there is a significant difference between these two games, which I did not discuss in section 3.5, where I assumed only Status Quo to be the initial state. If the initial state is (3,4) in game 35 – Magnanimity in MG – the players will not move from this state either, making it the second NME in game 35. By contrast, game 22a has no second NME: every initial state goes into (4,2).

The fact that one game (22a) has one NME, and the other game (35) two NMEs, is not exactly news, because I have already illustrated and

[4] This section and the next are adapted from Brams (forthcoming).

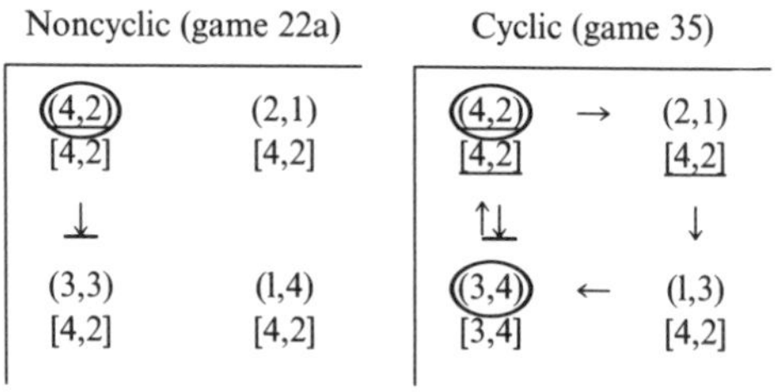

Noncyclic (game 22a)
Cyclic (game 35)
(4,2)
[4,2]
(2,1)
[4,2]
(3,3)
[4,2]
(1,4)
[4,2]
(4,2)
[4,2]
(2,1)
[4,2]
(3,4)
[3,4]
(1,3)
[4,2]

Key: (*x*,*y*) = (payoff to *V*, payoff to *D*)
[*x*,*y*] = [payoff to *V*, payoff to *D*] in anticipation game (AG)
4 = best; 3 = next best; 2 = next worst; 1 = worst
Nash equilibria in original games and AG of game 35 underscored
NMEs circled
Unblocked arrows indicate direction of cycling in game 35

Figure 4.1 Two class 1 Magnanimity Games (22a and 35)

interpreted several 2×2 strict ordinal games that have between one and three NMEs. What is news in this chapter is that game 35 is "cyclic," whereas game 22a is not.

To illustrate this distinction, first consider game 22 in figure 4.1.[5] Cycling will not occur – in either a clockwise or counterclockwise direction – in this game because moves from every state always bring the process to a state where the player who moves next receives its best payoff of 4, making a move from this state irrational, according to rule 5′. To wit,

- in a clockwise direction, the move by *R* from (2,1) to (1,4) gives *C* its best payoff, so *C* will not move from (1,4), as shown by the blocked arrow emanating from (1,4); and
- in a counterclockwise direction, the move by *C* from (2,1) to (4,2) gives *R* is best payoff, so *R* will not move from (4,2), as shown by the blocked arrow emanating from (4,2).

Now consider game 35 in figure 4.1. Although a counterclockwise move by *C* from (2,1) to (4,2) gives *R* its best payoff, preventing cycling in a counterclockwise direction – as shown by the blocked arrow emanating from (4,2) – moves in a clockwise direction never give a player its best payoff when it has the next move: *R* at (2,1), *C* at (1,3), *R* at (3,4), and *C* at (4,2) never receive payoffs of 4, making cycling in a clockwise direction possible, according to rule 5′, as shown by the clockwise arrows in figure 4.1.

[5] I temporarily drop the "a" from game 22a, because, as will become evident, the effects of moving power do not depend on whether play starts at (4,2) in game 22a, (4,1) in game 22b, or either of the other two states in game 22 in the Appendix.

The fact that clockwise moves around the payoff matrix of game 35 do not violate rule 5′ renders this game cyclic. On the other hand, game 22, in which cycling in both directions runs amok of a player receiving its best payoff when its turn to move comes up, is noncyclic.

There are 36 2×2 cyclic games, like game 35, in which blockage does not occur according to rule 5′. In these games, as I next show, cycling can occur in only one direction.

> **Theorem 4.1.** *If a 2×2 strict ordinal game is cyclic, cycling can occur either in a clockwise direction or in a counterclockwise direction but not in both directions.*

Proof. On the left-hand side of figure 4.2, I show the four placements of the best payoffs for *C* (4) that preclude cycling in either a clockwise direction (top matrix) or a counterclockwise direction (bottom matrix), as indicated by the blockages of the arrows. In the top left-hand matrix, for example, where *C* obtains a 4 at both the main-diagonal entries, cycling in a clockwise direction is irrational, because the process will stop when *C* obtains its best payoff, as shown by the blockages of the arrows emanating from these states. Likewise, in the bottom left-hand matrix where *C* obtains a 4 at both the off-diagonal entries, cycling in a counterclockwise direction is irrational, because the process will stop when *C* obtains its best payoff, again as shown by the blockages.

But *C* must obtain its best payoff in one of the four states of the two left-hand matrices, so one of the four entries must contain a 4. Depending on which entry, cycling either in a clockwise or in a counterclockwise direction, but not both, will be precluded. A similar argument holds for why clockwise and counterclockwise cycling, based on the placement of 4's for *R* in the right-hand matrices, cannot both be precluded.

Now the placement of a 4 by *C* and a 4 by *R* gives rise to three mutually exclusive and exhaustive possibilities:

(1) Both players' 4s prohibit clockwise cycling but not counterclockwise cycling;

(2) Both players' 4s prohibit counterclockwise cycling but not clockwise cycling;

(3) One player's 4 prohibits clockwise cycling, and the other player's 4 prohibits counterclockwise cycling, which makes the game noncyclic.

Whichever of (1), (2), or (3) obtains, cycling cannot occur in both directions at the same time. Q.E.D.

Define a game to be *symmetric* if there is an arrangement of the payoffs to the players so that the payoffs along the main diagonal are the same, whereas the off-diagonal payoffs are mirror images of each other (i.e., those for *C* and *R* are interchanged). Twelve of the 78 2×2 games are

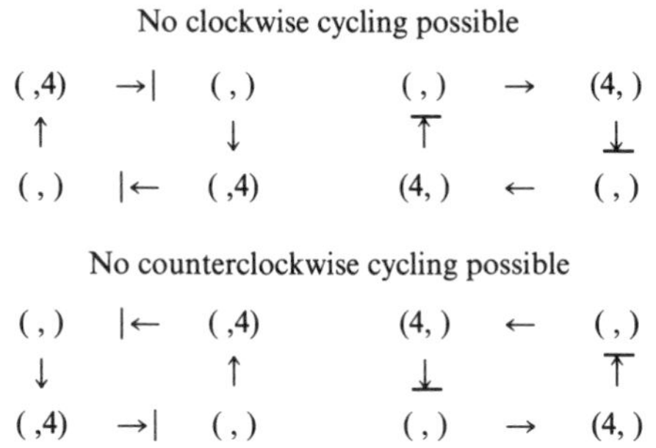

Figure 4.2 Cycling possibilities in a 2×2 game

symmetric, which are illustrated by Prisoners' Dilemma (game 32) and Chicken (game 57) in figure 3.5.

Corollary 4.1. *If a 2×2 strict ordinal game is symmetric, it is noncyclic.*

Proof. By Theorem 4.1, the payoffs of one player prohibit cycling in one direction. Because either the mirror-image off-diagonal payoff of the other player is a 4, which causes blockage in the other direction, or a main-diagonal payoff is (4,4), which causes blockage in both directions, a symmetric game does not cycle in either direction. Q.E.D.

To summarize, no symmetric game is cyclic; if an asymmetric game is cyclic, cycling can go in only one direction. I next divide the 36 cyclic games into different classes, depending on how much "friction" the players encounter in cycling.

A game may be cyclic either in a clockwise direction, in which case the 4s for *C* and *R* fit the pattern of the bottom two matrices in figure 4.2 (which preclude counterclockwise cycling), or in a counterclockwise direction, in which case the 4s for *C* and *R* fit the pattern of the top two matrices in figure 4.2 (which preclude clockwise cycling). The 36 cyclic games I now present all fit the latter pattern, so cycling always occurs in a counterclockwise direction.[6]

I next divide the 36 cyclic games into three mutually exclusive classes, depending on the number of "impediments" that players encounter in moving, in a counterclockwise direction, around the matrix:

[6] I have written the games so that cycling is counterclockwise principally because Rapoport and Guyer (1966) list the 36 cyclic games (which they did not identify as such) in this manner. To facilitate linking their taxonomy of the 78 2×2 strict ordinal games with the 36 that are cyclic, I include in parentheses, after the numbers I assign to the 36 cyclic games, the Rapoport–Guyer numbers; these are also given in the Appendix for the 57 2×2 conflict games in which there is no mutually best (4,4) state.

1 Moving power is *effective* – the outcome that each player can induce with moving power is better for it than the outcome that the other player can induce (6 games).

44 (75)

$(2,3)^c$	(4,1)
$(3,2)^r$	(1,4)

45 (76)

$(2,3)^c$	(3,1)
$(4,2)^r$	(1,4)

46 (70)

$(3,4)^c$	(2,1)
$(4,2)^r$	(1,3)

47 (71)

$(3,3)^c$	(2,1)
$(4,2)^r$	(1,4)

42 (73)

$(2,4)^c$	(4,1)
$(3,2)^r$	(1,3)

43 (74)

$(2,4)^c$	(3,1)
$(4,2)^r$	(1,3)

2 Moving power is *irrelevant* – the outcome induced by one player is better for both players (3 games).

29 (72)

$(3,2)^c$	(2,1)
$(4,3)^{r*}$	(1,4)

30 (77)

$(2,2)^c$	(4,1)
$(3,3)^{r*}$	(1,4)

31 (78)

$(2,2)^c$	(3,1)
$(4,3)^{r*}$	(1,4)

Key: (x,y) = (payoff to Row, payoff to Column)
4 = best; 3 = next best; 2 = next worst; 1 = worst
c = best state C can induce with moving power
r = best state R can induce with moving power
* = induced state better for both players

Figure 4.3 9 strongly cyclic games

1 No player has an impediment – strongly cyclic games. The nine games in this category are shown in figure 4.3.[7] At every state in these games, one player does immediately better by switching its strategy. In game 44, for example, assume the initial state is (2,3) in the upper left-hand cell. Then *R* does better moving to (3,2), whence *C* does better moving to (1,4), whence *R* does better moving to (4,1), whence *C* does better returning to state (2,3). In other words, there is no *impediment* to cycling: at every state a player does immediately better moving in a counterclockwise direction.

2 One player has an impediment – moderately cyclic games. The 18 games in this category are shown in figure 4.4. At one state in these games, which is always the upper left-hand cell in each matrix in figure 4.4, *R* always

[7] Ignore for now the *r* and *c* superscripts in these games and the games in figures 4.4 and 4.5. I shall explain both this notation and the classification of games based on moving power later.

1 Moving power is *effective* – the outcome that each player can induce with moving power is better for it than the outcome that the other player can induce (8 games).

36 (49)

$\underline{(3,4)}^c$	$(4,3)^r$
(2,1)	(1,2)

37 (50)

$\underline{(3,4)}^c$	$(4,3)^r$
(1,1)	(2,2)

38 (51)

$\underline{(3,4)}^c$	$(4,2)^r$
(2,1)	(1,3)

39 (52)

$\underline{(3,4)}^c$	$(4,2)^r$
(1,1)	(2,3)

40 (53)

$\underline{(3,3)}^c$	$(4,2)^r$
(2,1)	(1,4)

41 (54)

$\underline{(3,3)}^c$	$(4,2)^r$
(1,1)	(2,4)

56 (56)

$\underline{(2,4)}$	$(4,2)^r$
(1,1)	$(3,3)^c$

48 (57)

$\underline{(2,3)}$	$(4,2)^r$
(1,1)	$(3,4)^c$

2 Moving power is *irrelevant* – the outcome induced by one player is better for both players (8 games).

12 (40)

$\underline{(3,4)}^{c*}$	(4,1)
$(2,2)^r$	(1,3)

13 (41)

$\underline{(3,4)}^{c*}$	(4,1)
$(1,2)^r$	(2,3)

23 (42)

$\underline{(3,3)}^{c*}$	(4,1)
$(2,2)^r$	(1,4)

24 (43)

$\underline{(3,3)}^{c*}$	(4,1)
$(1,2)^r$	(2,4)

27 (47)

$\underline{(2,3)}$	(4,1)
$(1,2)^r$	$(3,4)^{c*}$

28 (48)

$\underline{(2,2)}$	(4,1)
$(1,3)^r$	$(3,4)^{c*}$

49 (44)

$\underline{(2,4)}$	(4,1)
$(1,2)^r$	$(3,3)^{c*}$

50 (55)

$\underline{(2,4)}$	$(4,3)^{r*}$
(1,1)	$(3,2)^c$

3 Moving power is *ineffective* – each player prefers the outcome that the other player can induce (2 games).

25 (45)

$\underline{(3,2)}^{c\#}$	(4,1)
$(2,3)^{r\#}$	(1,4)

26 (46)

$\underline{(3,2)}^{c\#}$	(4,1)
$(1,3)^{r\#}$	(2,4)

Key: (x,y) = (payoff to Row, payoff to Column)
4 = best; 3 = next best; 2 = next worst; 1 = worst
c = best state C can induce with moving power
r = best state R can induce with moving power
* = induced state better for both players
# = induced state of other player preferred
Nash equilibria underscored

Figure 4.4 18 moderately cyclic games

does worse moving to the lower left-hand cell. In game 36, for example, *R* does worse moving from (3,4) to (2,1). However, all subsequent moves in a counterclockwise direction are immediately beneficial to the mover: *C* in moving from (2,1) to (1,2), *R* in moving from (1,2) to (4,3), and *C* in moving from (4,3) back to (3,4).

3 Both players have an impediment – weakly cyclic games. The nine games in this category are shown in figure 4.5. At two states in these games, which are always the upper left-hand and the lower left-hand cells in each matrix, the players do worse moving in a counterclockwise direction. In game 33, for example, *R* does worse moving from (3,4) to (1,2), and *C* does worse moving from (1,2) to (2,1). Subsequently, however, *R* does better moving from (2,1) to (4,3), and *C* does better in returning to (3,4) from (4,3) to complete the cycle.

> **Theorem 4.2.** *There are no 2×2 cyclic games in which there are two separated impediments. In particular, there are no games in which there are three impediments.*

Proof. Two impediments are *separated* if, after one player does worse by moving, the same player does worse by moving again (i.e., after the second player has moved). Assume the first player is *R*, and it does worse moving in a counterclockwise direction on the two occasions when it has the opportunity to do so: from both the upper-left to the lower-left cell, and from the lower-right to the upper-right cell. Then it cannot receive 4 at either of the cells from which it moves, because otherwise the game would not cycle. But if *R* received 4 at either of the other two cells, then moves to these cells would necessarily be beneficial, which is contrary to the supposition that there is an impediment at both. The fact that there cannot be two separated impediments also proves that there cannot be three impediments, because the three would have to include the two that are separated. Q.E.D.

Theorem 4.2 establishes that the three classes of cyclic games – strongly cyclic (no friction), moderately cyclic (some friction), and weakly cyclic (much friction) – are exhaustive. Although the friction of one or two impediments would seem to impede cycling, I shall next show that even the two impediments of some weakly cyclic games can be overcome if one player has moving power.

4.3 Moving power

Assume, as before, that the players not only know their own payoffs but also have complete information about the payoffs of their opponents. If

1 Moving power is *effective* – the outcome that each player can induce with moving power is better for it than the outcome that the other player can induce (2 games).

33 (19)

$(3,4)^{c}$	$(4,3)^{r}$
(1,2)	(2,1)

34 (20)

$(3,4)^{c}$	$(4,3)^{r}$
(2,2)	(1,1)

2 Moving power is *irrelevant* – the outcome induced by one player is better for both players (5 games).

1 (13)

$(3,4)^{c*}$	(4,2)
$(2,3)^{r}$	(1,1)

2 (14)

$(3,4)^{c*}$	(4,2)
$(1,3)^{r}$	(2,1)

3 (15)

$(3,4)^{c*}$	(4,1)
$(2,3)^{r}$	(1,2)

4 (16)

$(3,4)^{c*}$	(4,1)
$(1,3)^{r}$	(2,2)

35 (21)

(2,4)	$(4,3)^{r*}$
(1,2)	$(3,1)^{c}$

3 Moving power is *ineffective* – each player prefers the outcome that the other player can induce (2 games).

5 (17)

(2,4)	(4,2)
$(1,3)^{r\#}$	$(3,1)^{c\#}$

6 (18)

(2,4)	(4,1)
$(1,3)^{r\#}$	$(3,2)^{c\#}$

Key: (x,y) = (payoff to Row, payoff to Column)
4 = best; 3 = next best; 2 = next worst; 1 = worst
c = best state C can induce with moving power
r = best state R can induce with moving power
* = induced state better for both players
\# = induced state of other player preferred
Nash equilibria underscored

Figure 4.5 9 weakly cyclic games

this is the case, when would a player have an incentive to cycle to try to outlast an opponent?

By "outlasting" an opponent, I mean that one (stronger) player can force the other (weaker) player to stop the move–countermove process at a state where the weaker player has the next move. Forcing stoppage at such a state, however, may not always lead to an outcome favorable to the stronger player. Not only may a stoppage-forcing strategy be unproductive, but it may actually be counterproductive, leading to a worse outcome than if the stronger player had stopped the cycling itself.

Rule 5′ specified what players would not do – namely, move from a best (4) state when it was their turn to move. By precluding cycling in both directions in 42 of the 78 games, however, this rule did not say anything about *where* cycling would stop in the remaining 36 cyclic games. In these games, a final rationality rule, which replaces rule 6 in section 1.4, is needed to say at what, if any, state play will terminate:

6′ In a cyclic game of complete information, *P*1 will move from an initial state, even if play returns to this state and repeatedly cycles, if it
 (i) has "moving power"; and
 (ii) can induce a better outcome for itself with this power.

*P*1 has *moving power* if it can induce *P*2 eventually to stop, in the process of cycling, at one of the two states at which *P*2 has the next move. The state at which *P*2 stops, I assume, is that which *P*2 prefers.[8]

I next describe and illustrate with examples three possible effects of moving power in cyclic games:

1 Moving power is *effective* – the outcome that each player can induce with moving power is better for it than the outcome that the other player can induce (16 games). To illustrate this effect, consider game 44 in figure 4.3, which, as noted in section 2.3 (see figure 2.3), is a game of total conflict: what is best (4) for one player is worst (1) for the other, and what is next best (3) for one is next worst (2) for the other. As I showed in section 4.2, game 44 is a strongly cyclic game, with cycling going in a counterclockwise direction.

[8] An earlier version of this concept was proposed in Brams (1982, 1983). Other concepts of power, based on the framework of TOM but not the rules given here, include threat power (Brams, 1983, 1990; Brams and Hessel, 1984), staying power (Brams, 1983; Brams and Hessel, 1983; Kilgour, De, and Hipel, 1987), and holding power (Kilgour and Zagare, 1987), but none of these is relevant to the study of cycles. (Threat power, and a new power concept, order power, are defined and analyzed in chapter 5.) Neither are the empirical studies based on TOM, including Brams (1985a, 1985b) and Zagare (1981, 1983, 1987). On the other hand, cycling is allowed but not assumed always to occur (as here) in De, Hipel, and Kilgour (1990), who propose a notion of "hierarchical power." Langlois (1992) permits cycling within a cardinal framework, rooted in expected-utility calculations; because cycling is assumed to be costly, however, both players will eventually want to desist. I do not make that assumption here; instead, I assume that the player with moving power is essentially indefatigable – at least compared to the player without moving power, who must eventually stop moving. (Implicitly, of course, this assumption suggests that cycling is costly, but without including cardinal utilities in the model, I cannot specify costs; thus I say that, whatever the costs, the player with moving power can better withstand them.) I also assume that only one player can possess moving power at any one time; otherwise, each player could force the other to terminate play at the same time – perhaps out of mutual exhaustion – which would indicate the lack of a power asymmetry.

Now assume that *R* possesses moving power in game 44. Because cycling is counterclockwise, *R* can induce *C* to stop at either (3,2) or (4,1), where *C* has the next move. Obviously, *C* would prefer (3,2), which is indicated as the moving-power outcome that *R* can induce by the superscript of *r*; it gives *R* its next-best state of 3 and *C* its next-worst state of 2.

On the other hand, if *C* possesses moving power, it can induce *R* to stop at either (2,3) or (1,4), where *R* has the next move. Obviously, *R* would prefer (2,3), which is indicated as the moving-power outcome that *C* can induce by the superscript of *c*; it gives *C* its next-best state of 3 and *R* its next-worst state of 2. In other words, the player with moving power can ensure a better outcome (3) than the player without (2), which makes this power *effective*: it is better for a player to possess moving power than for the other player to possess it.

A glance at figures 4.3, 4.4, and 4.5 shows that every class of cyclic games (strongly cyclic, moderately cyclic, and weakly cyclic) includes some games in which moving power is effective. The different outcomes that *R* and *C* can induce are indicated by the superscripts of *r* and *c*. Notice that *R* always prefers the outcome it can induce to the outcome that *C* can induce, and vice-versa, in games in which moving power is effective.

2 Moving power is *irrelevant* – the outcome induced by one player is better for both players (16 games). This category of games includes those in which both players prefer the (Pareto-optimal) outcome that one player can induce to the (Pareto-inferior) outcome that the other player can induce with moving power. The preferred outcome, which is indicated by a "*" next to *r* or *c*, is therefore the one where the move–countermove process will stop: because neither player can induce a better outcome for itself – independent of which player has moving power – moving power is *irrelevant*.

Game 29 in figure 4.3 is an example of a game in which moving power is irrelevant. If *R* possesses moving power, it can induce *C* to stop at either (4,3) or (2,1), where *C* has the next move. Clearly, *C* would prefer (4,3). If *C* possesses moving power, it can induce *R* to stop at either (3,2) or (1,4), where *R* has the next move. Clearly, *R* would prefer (3,2). However, the fact that *C* prefers (4,3) to (3,2) means that it would accede to *R*'s moving-power outcome, (4,3), which is starred, rather than induce its own, (3,2).

Observe that, moving counterclockwise, *C* has the next move at (4,3) in game 29, so it would be in its interest to stop the process there without being forced to stop by *R*. Indeed, it would be counterproductive for *C*, if it possessed moving power, to force *R* to stop at (3,2). Thus, moving power in game 29 is irrelevant – (4,3) would be the outcome to which both players would be drawn, whichever player possessed moving power. Every class of cyclic games, as shown in figures 4.3, 4.4, and 4.5, contains some in which moving power is irrelevant.

3 Moving power is *ineffective* – each player prefers the outcome that the other player can induce (4 games). Paradoxically, an opponent's exercise of moving power in a few games leads to a better outcome than a player can induce for itself. Because this is the opposite of the situation when moving power is effective, I call moving power *ineffective* in such games and indicate it by "#" next to *r* and *c*.

As an example, consider game 25 in figure 4.4. Like game 44 in figure 4.3, this is a game of total conflict. Here, however, the best outcome that *R* can induce with moving power is (2,3), because *C* prefers this state to the other state where it can stop, (4,1), when it has the next move. Similarly, the best outcome that *C* can induce with moving power is (3,2), because *R* prefers this state to the other state where it can stop, (1,4), when it has the next move. Thus, although each player can induce an outcome that yields its next-worst payoff of 2, each would prefer the outcome that its opponent can induce, which yields its next-best payoff of 3.

Notice that the outcome that *R* can induce in all the other power-ineffective games (games 26, 5, and 6) is (1,3), whereas *C* can induce (3,1) in game 5. Not only are these the worst outcomes for the inducers, but *R* can always obtain a better payoff of 2 by itself stopping at (2,4) in games 26, 5, and 6, and *C* by stopping at (4,2) in game 5.

Even at the latter states, however, the player who stops does worse than at the outcomes its opponent can induce, either by inducing stoppage (3) or by stopping itself (4). Thus, the possession of moving power in these games does not benefit either player. Moreover, unlike the games in which moving power is irrelevant, there is no single (Pareto-optimal) outcome that one of the players can induce that both would prefer to the (Pareto-inferior) outcome that the other player can induce.

It turns out that moving power is never ineffective in strongly cyclic games (figure 4.3). However, there are both moderately and weakly cyclic games in which the possession of moving power can lead to an outcome inferior to one that would occur if the other player possessed moving power and exercised it (figures 4.4 and 4.5).

The fact that moving power is of no help to players in power-ineffective games renders the use of the word "power" – which normally connotes an ability to induce better outcomes than one could obtain without it –

dubious in these games. I next state three propositions that summarize properties of games in which moving power is effective or irrelevant:

Proposition 4.1. *In the two games that have Pareto-inferior Nash equilibria and in which moving power is irrelevant (games 27 and 28 in figure 4.4), the outcome induced by cycling is Pareto-superior to the Nash equilibrium.*

Thus, it is not necessary that there be an asymmetry in capabilities in order for players to achieve an outcome that classical game theory considers unstable (i.e., not a Nash equilibrium). Contemplating the possibility of cycling according to rules 5′ and 6′ in games 27 and 28, the players would realize that they have no incentive to stay at the unique Nash equilibrium in these games, because they both prefer the unique state Pareto-superior to it. Hence, even though moving power is irrelevant in these games, cycling singles out the latter state as the one where both players would find it rational to terminate the cycling.

The counterpart to Proposition 4.1, when moving power is effective, is the following:

Proposition 4.2. *In the one game that has a Pareto-inferior Nash equilibrium and in which moving power is effective (game 48 in figure 4.4), one of the players can induce an outcome Pareto-superior to the Nash equilibrium.*

In game 48 in figure 4.4, (2,3) is the Pareto-inferior Nash equilibrium. If *C* has moving power, it can induce the Pareto-superior outcome, (3,4), whereas *R* can induce (4,2), which is Pareto-optimal but is not Pareto-superior to (2,3). In either event, a player's possession of moving power obviates the choice of (2,3).

Apart from Nash equilibria, Pareto-optimality can always be achieved:

Proposition 4.3. *In all games in which moving power is effective or irrelevant, the induced outcomes (two when moving power is effective, one when it is irrelevant) are Pareto-optimal.*

Thus, in the vast majority of cyclic games (89 percent, excluding only the four power-ineffective games), a Pareto-optimal outcome can be expected, whichever player possesses moving power. Even in the power-ineffective games (games 25, 26, 5, and 6), a Pareto-optimal outcome seems likely unless a player foolishly tries to exercise moving power when it could do better by stopping play itself. Of course, these results on Pareto-optimal outcomes depend on both players behaving according to rules 5′ and 6′, rather than rules 5 and 6 that prohibit cycling.

One way to characterize the difference between these two sets of rationality rules is in terms of whether induction is "backward" or "forward." The backward induction of rules 5 and 6 assumes an anchor of exactly four moves ahead, after the completion of one cycle. As I showed in section 1.4, players decide, working backwards from this anchor, where to stop before completing a cycle.[9]

By contrast, induction is *forward* if a player indicates at the start of play that it is willing and able to cycle indefinitely to assert its moving power. It may do this by continuing to move if an opponent does, thereby signaling that it "means business." But this display of resoluteness does not mean that it will in fact be able to outlast its opponent in repeated cycling, because which player has moving power may not be *common knowledge* – that is, known to both players, with each knowing that the other knows, knowing that the other knows that each knows, and so on *ad infinitum.*

Unlike backward induction, forward induction has no anchor, just repeated cycling – or the threat of it – an indefinite number of times. Because of this indefiniteness, players do not reason backward but must, instead, look forward, calculating who must eventually stop, or testing each other on this point if information is incomplete.[10]

Not surprisingly, the backward-looking and forward-looking approaches to nonmyopic calculation may give, on occasion, different results. Which is the appropriate kind of induction to apply depends, ultimately, on the nature of the situation being analyzed, including whether one player possesses moving power and can effectively use it.

If this is not the case, the players may want to end their conflict before cycling, which is the assumption of rules 5 and 6. Likewise, in games in which moving power is irrelevant, one would not expect the players to cycle indefinitely; instead, they would stop at the Pareto-optimal outcome that one player can induce, and both players prefer, to the Pareto-inferior outcome that the other player can induce.

On the other hand, if one player thinks it has greater stamina and, moreover, moving power is effective in the game being played, it will want to continue to move, according to rules 5′ and 6′, if its claim to this power is disputed by the other player. But in games in which moving power is effective, would players physically cycle in order to implement their preferred outcomes, or are their moves likely to be mental ones, based on a thought experiment?

There is no reason why rational players would make physical moves in power-effective games if there is common knowledge about which player has moving power. In this case, it would be rational for the weaker player

9 Recall that if it is rational, according to backward induction, to complete a cycle, players will not depart from the initial state in the first place.

10 This distinction between backward and forward induction is different from that customarily made in the game-theoretic literature. For the conventional distinction, see Myerson (1991), Fudenberg and Tirole (1991), and Binmore (1992).

to acquiesce in the stronger player's moving-power outcome, immediately terminating play when this state is reached.

But in many real-world conflicts – as well as other-world (e.g., theological) conflicts, as I shall argue in section 4.4 – there may be no clear recognition of which, if either, player has moving power. In fact, there may be a good deal of misinformation. For example, if both players believe they can hold out longer, cycling is likely to persist until one player succeeds in demonstrating its greater strength, or both players are exhausted by the repeated cycling.

This seems to have occurred in the Egyptian–Israeli conflict, but it still required great pressure from the United States to bring about a peaceful resolution. As I indicated in section 4.1, if this conflict can be modeled by a class 2 MG game, in which Magnanimity is the unique NME, it should not have taken 31 years to resolve. Thus, I believe, the resolution probably required the demonstration of moving power by Egypt, whose possession of it, as *D*, enables it to be effective in inducing Magnanimity in three of the four class 1 games in figure 3.5 (namely, games 33, 34, and 35). Only game 22a in class 1 is impervious to *D*'s moving power.

One or more of the former three games may well represent the Egyptian–Israeli conflict, and the preferences of the players, in the aftermaths of each war from 1948 to 1973. If so, they, like game 22a, all have Status Quo as the NME when play commences after a war.

In MG games 33, 34, and 35, Israel has no incentive to budge from Status Quo – unless Egypt demonstrates its moving power. Once it has, however, Magnanimity becomes the rational outcome, giving Israel its next-best and Egypt its best payoffs of (3,4) in games 33, 34, and 35.

In game 22a, by contrast, Magnanimity gives payoffs of (3,3) to the players. But because this game is noncyclic, Magnanimity cannot be induced by Egypt's moving power. Instead, Status Quo of (4,2), because it is the unique NME (and Nash equilibrium), is the indisputably rational outcome, which favors Israel.

Game 22a, however, is an implausible representation of the Egyptian–Israeli conflict. Thus, at Rejected Magnanimity of (1,4) in this game, Egypt obtains its best payoff and Israel its worst. But why would Egypt do better rejecting concessions from Israel? In my opinion, not only is this an inaccurate description of Egypt's preferences in 1978–9, but games 33, 34, and 35 all seem closer to the mark. Unlike game 22a, for example, both sides in these games suffer a Pareto-inferior state whenever *D* (Egypt) chooses $\bar{C}$, whichever strategy *V* (Israel) chooses. Furthermore, all these games result in Magnanimity, not Status Quo, when Israel recognizes Egypt's moving power and acts accordingly.

But this is not to say that Egypt "won" at Camp David in 1978; on the contrary, Quandt (1986, pp. 254–8) argues that Israel, comparatively speaking, wrung more concessions out of Egypt than vice versa. Nevertheless, because of the nature of the bargaining process, both sides probably went to their "bottom lines" (Brams, 1990, pp. 57–60).

They did so in part because Magnanimity for both players was better in 1978–9 than Rejected Magnanimity, which is true in every class 1 MG game except 22a. Consequently, I think one of the other three class 1 games was played after 1973, rendering Magnanimity rational when Israel, in the end, could do no better than accept Egypt's moving power.

I next will give an example of a cyclic game in which each player can induce its best state if it possesses moving power, making this power effective. (In the specific MG games just considered, moving power is effective in games 33 and 34, irrelevant in game 35, and not defined in game 22a because this game is noncyclic.) I suppose, in my initial interpretation, that only one player (a superior being) possesses this power.

4.4 The Revelation Game

Using a 2×2 game to model the relationship that a person (P) might have with a superior being (SB), like God, drastically simplifies a deep and profound religious experience for many people.[11] My aim, however, is not to describe this experience but to abstract from it, using the game to analyze a central theological question: Can belief in SB be conceptualized as a rational choice?

The answer depends, in part, on whether it is proper to view SB as a game player, capable, like P, of making independent choices. Or is SB too ethereal or metaphysical an entity to depict in these terms? Consider the view expressed by the theologian, Martin Buber (2nd edn, 1958, p. 135), about his approach to understanding God:

> The description of God as a Person is indispensable for everyone who like myself means by "God" not a principle ... not an idea ... but who rather means by "God," as I do, him who – whatever else he may be – enters into a direct relation with us ...

It is not a great leap of faith, in my view, to model a "direct relation" as a game.[12]

The game I shall use to explore the rationality of belief in an SB is the

[11] This section is adapted from Brams (1983, pp. 15–24, 101–4, 145) with permission; all translations are from *The Torah: The Five Books of Moses* (2nd edn, 1967).

[12] As Cohen (1991, p. 24) points out, however, in the non-Western world "the concept of a personal, unmediated relationship between human being and deity is quite incomprehensible."

Revelation Game (RG), which supposes specific goals of P and SB. To preview the subsequent analysis, I will show that

- play of this game leads to a Pareto-inferior outcome, based on the standard theory; but
- both P and SB can induce Pareto-optimal outcomes in this game if one or the other possesses moving power.

I will also discuss linkages between the assumptions of the game and the Hebrew Bible and comment on how the players in RG might behave cyclically over time.

In RG, I assume that SB has two strategies: reveal itself (R), which establishes its existence, and don't reveal itself ($\bar{R}$), which does not establish its existence. Similarly, P has two strategies: believe in SB's existence (B), and do not believe in SB's existence ($\bar{B}$).

Instead of writing down and trying to justify the preferences of the players at each state, as I have done until now, I begin by specifying (i) primary and (ii) secondary goals of each player in RG:[13]

SB: (i) Wants P to believe in its existence;
(ii) Prefers not to reveal itself.

P: (i) Wants belief (or nonbelief) in SB's existence confirmed by evidence (or lack thereof);
(ii) Prefers to believe in SB's existence.

The primary and secondary goals of each player, taken together, completely specify their orderings of states from best to worst. The primary goal distinguishes between the two best (4 and 3) and the two worst (2 and 1) states of a player, whereas the secondary goal distinguishes between 4 and 3, on the one hand, and 2 and 1 on the other.[14]

Thus for SB, (i) establishes that it prefers states in the first column (4 and 3) of the figure 4.6 matrix (associated with P's strategy of B) to states in the second column (2 and 1) of the matrix (associated with P's strategy of $\bar{B}$). Between the two states in each column, (ii) establishes that SB prefers not to reveal itself (hence, 4 and 2 are associated with over $\bar{R}$) to revealing itself (3 and 1 are associated with R).

Likewise for P, (i) says that it prefers to have its belief or nonbelief confirmed by evidence (so the main-diagonal states are 4 and 3) to being unconfirmed (so the off-diagonal states are 2 and 1). Between the pairs of main-diagonal and off-diagonal states, (ii) says that P prefers to believe (so 4 and 2 are associated with B) rather than not to believe (so 3 and 1 are

[13] True, I also took this approach in the case of the mugging game in section 3.2 (figure 3.2), even proposing a tertiary goal for each player. However, I did not formally derive the payoff matrix of this game from the postulated goals, which I shall do in the case of RG.

[14] This is an example of a *lexicographic decision rule*, whereby states are first ordered on the basis of a most important criterion, then a next most important criterion, and so on (Fishburn, 1974).

	P			
SB	Believe in *SB*'s existence (*B*)		Don't believe in *SB*'s existence ($\bar{B}$)	
Reveal itself (*R*) (establish its existence)	I *P* faithful with evidence: belief in existence confirmed (3,4)[c] [3,4]	←	II *P* unfaithful despite evidence: nonbelief in existence unconfirmed (1,1) [3,4]	
	↓		↑	
Don't reveal itself ($\bar{R}$) (don't establish its existence)	IV *P* faithful without evidence: belief in existence unconfirmed (4,2)[r] [4,2]	→	III *P* unfaithful without evidence: nonbelief in existence confirmed (2,3) [3,4]	←Dominant strategy

Key: (*x,y*) = (payoff to *SB*, payoff to *P*)
[*x,y*] = [payoff to *SB*, payoff to *P*] in anticipation game (AG)
4 = best; 3 = next best; 2 = next worst; 1 = worst
c = state Column (*P*) can induce with moving power
r = state Row (*SB*) can induce with moving power
Nash equilibria in original game and AG underscored
NMEs circled
Arrows indicate direction of cycling

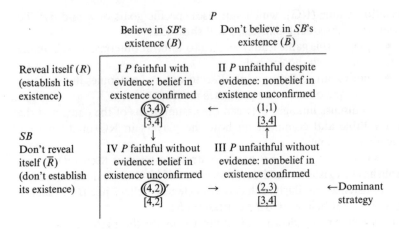

Figure 4.6 Revelation Game (game 48)

associated with $\bar{B}$. The game so defined by the goals of both players is moderately cyclic game 48 (the row and column strategies of the two players in its figure 4.4 representation are interchanged in figure 4.6), in which power is effective.

In the contemporary world, I would submit, evidence from one's observations, experiences, and reflections accumulates that predisposes one to believe or not believe in the existence of God or some other supernatural being or force – or leaves the issue open. How beliefs are formed about a deity is less well understood.[15]

Of course, religions predispose one toward particular views, and religious works may reinforce them. I next offer some brief remarks on the Hebrew Bible, which may lend plausibility to the goals of *P* and *SB* that I have assumed.

Evidence that the biblical God wanted His supremacy acknowledged by both Israelites and non-Israelites is plentiful in the Hebrew Bible.

15 For a developmental analysis of faith, see Fowler (1981); different kinds of theological evidence, and the different kinds of rationality that they give rise to, are discussed in Swinburne (1981, chapters 2 and 3).

Moreover, the biblical narratives make plain that He pursued this goal with a vengeance not only by severely punishing those who did not adhere to His commands and precepts but also by bestowing rewards on the faithful who demonstrated their unswerving belief through good deeds and sacrifices.

Yet beyond providing indirect evidence of His presence through displays of His might and miraculous powers, the biblical God had an overarching reason for not revealing Himself directly: it would undermine any true test of a person's faith, which I assume to be belief in God *not* necessarily corroborated by direct evidence. Only to Moses did God confirm His existence directly – "face-to-face" (Exodus 33:11; Numbers 12:6–8; Deuteronomy 34:10) – but that Moses actually saw God firsthand is contradicted by the statement God made to Moses: "But," He said, "you cannot see My face, for man may not see Me and live" (Exodus 33:20).

Because a person cannot be truly tested if God's existence has already been confirmed by some unequivocal revelatory experience, I assume God most desires from His subjects an expression of belief that relies only on faith (i.e., belief without direct evidence). Indeed, it is not unfair, in my opinion, to read the Bible as the almost obsessive testing of human beings by God to distinguish the faithful from those whose commitment to Him is lacking in zeal or persistence (remember that Job's faith faltered, but he never abandoned God).

This all-too-brief justification of *SB*'s goals by way of the biblical God's statements and actions will not be persuasive to those who regard the Bible as an unreliable source at best, pure fantasy at worst.[16] It is not, however, a nonbeliever – or, for that matter, a believer – whom I postulate as *P* in RG. Instead, I assume that *P* is somebody who takes the Bible (or other monotheistic religious works) seriously. Although these works may describe experiences that are outside *P*'s ken or beyond the secular world, I suppose that *P* has yet to make up its mind about the existence of an "ultimate reality" embodied in some *SB*.

While *P* entertains the possibility of *SB*'s existence, and in fact would prefer confirmatory to nonconfirmatory evidence in RG (according to its secondary goal), evidence is *P*'s major concern (i.e., its primary goal). Moreover, *P* realizes that whether or not *SB* provides it will depend on what *SB*'s rational choice in RG is.

To highlight the quandary that RG poses for both players according to the standard theory, observe that *SB* has a dominant strategy of $\bar{R}$: this strategy is better for *SB* whether *P* selects *B* [because *SB* prefers (4,2) to

16 For more evidence on God's goals beyond the cursory biblical citations provided here, see Brams (1980).

(3,4)] or $\bar{B}$ [because *SB* prefers (2,3) to (1,1)]. Given *SB*s dominant strategy of $\bar{R}$, *P*, which does not have a dominant strategy but prefers (2,3) to (4,2) in the second row of RG, will choose $\bar{B}$ as a best response. These strategies lead to the selection of (2,3), which is the unique Nash equilibrium in RG but Pareto-inferior to (3,4).

Even though (3,4) is better for both players than (2,3), (3,4) is not a Nash equilibrium because *SB* has an incentive, once at (3,4), to depart to (4,2). But neither is (4,2) an equilibrium, because once there *P* would prefer to move to (2,3). As I indicated earlier, RG is moderately cyclic, moving in a counterclockwise direction, as shown by the arrows in figure 4.6, with a single impediment occurring when *R* moves from (2,3) to (1,1).

Before applying TOM – and, specifically, the concept of moving power – to RG, let me clarify *SB*'s choice of $\bar{R}$, which I interpreted earlier as "don't establish its existence" (see figure 4.6). From *P*'s perspective, $\bar{R}$ may occur for two distinct reasons: (1) *SB* does not in fact exist, or (2) *SB* does not choose to reveal itself. Not only can *P* not distinguish between these two reasons for nonrevelation, but even if *SB* exists, *P* knows that *SB* has a dominant strategy of $\bar{R}$ and would, therefore, presumably choose it in RG.

For this reason, I do not assume that *P* would ever think there is conclusive evidence of nonexistence, so I do not give *P* this option in RG. Instead, *P* can choose not to believe in *SB*'s existence *and* – though this is not shown in the matrix – not to believe in *SB*'s nonexistence, either, which is to say that *P* is agnostic. That is, *P* suspends judgment, which I interpret as a kind of commitment to remain noncommittal.[17]

In a sense, a thoughtful agnostic plays RG all its life, never certain about *SB*'s strategy choice, or even that *SB* exists. In choosing $\bar{B}$, I interpret *P* to be saying that it does not believe either in *SB*'s existence or nonexistence *yet* – in other words, it wants to keep its options open.

Should *P* become a believer or a nonbeliever, then it no longer would be torn by the self-doubt reflected in its choices in RG. The evidence, so to speak, would be in. But I assume that *P* is neither an avowed theist nor an avowed atheist but a person with a scientific bent, who desires confirmation of either belief or nonbelief. Preferring the former to the latter as a secondary goal, *P* is definitely not an inveterate skeptic.

What *SB* might desire, on the other hand, is harder to discern. Certainly the God of the Hebrew Bible very much sought, especially from His

[17] Not everyone believes such openness is desirable, at least in the case of God. For example, Hanson (1971, pp. 303–31) thinks that the proper position of the agnostic on the question of God's existence should be one of reasonable doubt. For Hanson, moreover, the evidence is tipped decisively against God's existence. I discuss the views of other thinkers about the rationality of believing in God (e.g., Blaise Pascal, who proposed a famous wager based on the probability that God exists) in Brams (1983).

chosen people, the Israelites, untrammeled faith and demonstrations of it. Although He never revealed Himself in any physical form, except possibly to Moses before he died, He continually demonstrated His powers in other ways, especially by punishing those he considered transgressors.

If *SB* has moving power and RG is played according to rules 5′ and 6′, *SB* can induce *P* to stop at either (4,2) and (1,1), where *P* has the next move. *P* would obviously prefer (4,2), which gives *SB* its best payoff: *P*'s belief without evidence satisfies both of *SB*'s goals. But *P* obtains only its next-worst payoff in this state; it satisfies only its secondary goal of believing, but not its primary goal of having evidence to support this belief.

Endowing *SB* with moving power raises a feasibility question (section 1.6). Whenever *P* oscillates between belief and nonbelief, I assume that *SB* can switch back and forth between revelation and nonrevelation. But once *SB* has established its existence, can it be denied? I suggest that this is possible, but only if one views RG as a game played out over a long period of time.

To illustrate this point, consider the situation recounted in chapter 19 of Exodus. After God "called Moses to the top of the mountain" (Exodus 19:20) to give him the ten commandments, there was "thunder and lightning, and a dense cloud ... and a very loud blast of the horn" (Exodus 19:16). This display provided strong evidence of God's existence to the Israelites, but for readers of the Bible today it is perhaps not so riveting.

Yet even the Israelites became wary and restive after Moses' absence on Mount Sinai for forty days and nights. With the complicity of Aaron, Moses' brother, they revolted and built themselves a golden calf. God's earlier displays of might and prowess had lost their immediacy and therefore their force.

This insurrection enraged Moses and God. Moses destroyed the ten commandments and, with God's assistance, provoked the slaughter of 3,000 Levites (a tribe of Israelites) for their idolatry.

Moving to the present, the basis of belief would seem even more fragile. Many people seek a more immediate revelatory experience than reading the Bible, and some find it. For those who do not, God remains hidden or beyond belief unless they can apprehend Him in other ways.

This is where the problem of revelation arises. Without a personal revelatory experience, or the reinforcement of one's belief in God that may come from reading the Bible or going to religious services, belief in God's existence may be difficult to sustain with unswerving faith.

Revelation, also, may be a matter of degree. If God appears with sound and fury, as He did at Mount Sinai, He may likewise disappear like the morning fog as memories of Him slowly fade. Thereby seeds of doubt are planted. But a renewal of faith may also occur if a person experiences some sort of spiritual awakening.

A wavering between belief and nonbelief created by *SB*'s moving between revelation and nonrevelation shows that *P*'s belief in *SB* may have a rational basis for being unstable. Sometimes the evidence manifests itself, sometimes not, in this cyclic game. What is significant in RG is that *SB*'s exercise of moving power, according to rules 5′ and 6′, is consistent with *SB*'s sporadic appearance and disappearance – and with *P*'s responding to revelation by belief, to nonrevelation by nonbelief (up to a point).

In the Bible, God seems to want to remain inscrutable, as the following colloquy suggests:

> Moses said to God, "When I come to the Israelites and say to them 'The God of your fathers has sent me to you,' and they ask me, 'What is His name?' what shall I say to them?" And God said to Moses, "Ehyeh-Asher-Ehyet" ["I Am That I Am"]. He continued, "Thus shall you say to the Israelites, 'Ehyey [I Am] sent me to you.'" (Exodus 3:13–14)

As enigmatic as this reply is, however, God is also quick to trumpet His deeds and demonstrate His powers, as I showed in section 2.4 in discussing the pursuit of the Israelites by Pharaoh.

Relying on faith alone, when reason dictates that it may be insufficient to sustain belief, produces an obvious tension in *P*. Over a lifetime, *P* may move back and forth between belief and nonbelief as seeming evidence appears and disappears. For example, the indescribable tragedy of the Holocaust destroyed the faith of many believers, especially Jews, in a benevolent God, and for some it will never be restored.

But for others it has been. Furthermore, many former nonbelievers have conversion experiences – sometimes induced by mystical episodes – and, as a result, pledge their lives to Christ or God. For still others, there is a more gradual drift either toward or away from religion and belief in an *SB*, which is often related to age.

More broadly, there are periods of religious revival and decline, which extend over generations and even centuries, that may reflect a collective consciousness about the presence or absence of an *SB* – or maybe both.[18]

[18] These may be akin to the "long cycles" discussed in the literature of international relations (Rosecrance, 1987; Goldstein, 1988; Thompson, 1992), whose statistical foundations and empirical validity have been vigorously debated (Beck, 1991; Goldstein, 1991). A wide-ranging review of the literature on longitudinal patterns (e.g., Kuznets cycles and Kondratieff waves) in economics and politics is given in Berry (1991), who concludes that there is strong evidence of cyclical behavior, just as Modis (1992) finds evidence for the logistic S-curve. My primary interest here is not in identifying patterns

As Kolakowski (1982, p. 140) remarked, "The world manifests God and conceals Him at the same time."

It is, of course, impossible to say whether an *SB*, behind the scenes, is ingeniously plotting its moves in response to the moves, in one direction or another, of individuals or of society. But this is not the first Age of Reason, though it has had different names in the past (e.g., Age of Enlightenment), in which people seek out a rational explanation. Nor will it be the last, probably again alternating with periods of religious reawakening (e.g., as occurred during the Crusades) that will also come and go. This ebb and flow is inherent in the instability of moves in RG, even if an *SB*, possessed of moving power, has its way on occasion and is able temporarily to implement (4,2).

Perhaps the principal difficulty for *SB* in making this outcome stick is that peoples' memories erode after a prolonged period of nonrevelation. Consequently, the foundations that support belief may crumble. Nonbelief sets up the need for some new revelatory experiences, sometimes embodied in a latter-day messiah, followed by a rise and then another collapse of faith.

If *P* is assumed to be the player who possesses moving power, then it can induce (3,4), which *SB* would prefer to (2,3), given that *SB* must stop at one or the other of these two states when it has the next move. If the idea of "forcing" *SB* to reveal itself – and, on this basis, for *P* to believe – sounds absurd, it is useful to recall that the biblical God exerted Himself mightily on occasion to demonstrate His awesome powers to new generations. By the same token, God left the stage at times in order to test a new generation's faith, usually being forced to return in order to try to foster belief again.

The effects of moving power, whether possessed by *SB* or *P*, seem best interpreted in RG as occurring over extended periods of time. Memories fade, inducing *SB* to move from nonrevelation to revelation when the next generation does not understand or appreciate *SB*'s earlier presence. Even when *SB* moves in the opposite direction, going from revelation to nonrevelation, its actions may not appear inconsistent if *P*, effectively, is a different player. Thereby the earlier concern I raised about infeasible moves is dissipated in an extended game in which the identity of *P* changes.

Because RG is a cyclic game with two NMEs, one of which each player can induce, it seems best viewed as a game of movement, in which either player, if it possesses moving power, can induce its best state. Yet this is usually only a temporary "passing through," because the other player

inductively, as is done in this literature, but rather in developing a deductive rational-choice model that explains cycling, either in the short or the long term.

will, according to rules 5′ and 6′, respond by switching states. Finally, the player without moving power will be forced to desist. But if this player is *P*, and it believes for a time without evidence, then eventually it will be replaced by another *P* that feels less piety in the face of an ineffable *SB*.

Feasibility may militate against too quick switches on the part of the players, but fundamentally RG is a game for the ages. Rules 5′ and 6′ seem more apt in this game than rules 5 and 6, and RG's fluidity – rather than the stability of NMEs that arise from each state – seems its most striking feature.

4.5 Bombing campaigns in Vietnam

On December 18, 1972, President Richard Nixon ordered an all-out bombing campaign against North Vietnam.[19] "Linebacker II," as it was called, was an attempt to force the North Vietnamese seriously to negotiate an end to the Vietnam war. Nixon's decision encountered severe criticism and "stirred up a great furor amongst the anti-war elements in Congress and in the public" (Sharp, 1978, p. 252).

It was not the first time that the United States had attempted to force the North Vietnamese to the negotiation table through bombing. From 1965 to 1968, President Lyndon Johnson oversaw a bombing campaign dubbed "Rolling Thunder," often choosing targets himself. It was rooted in the idea that

> carefully calculated doses of force could bring about predictable and desirable responses from Hanoi. The threat implicit in minimum but slowly increasing amounts of force ... would, it was hoped by some, ultimately bring Hanoi to the table on terms favorable to the U.S. (Sharp, 1978, pp. 52–3)

But the campaign, gradualist in nature, failed miserably in its goal (Thompson, 1980). As early as 1965, the outgoing director of the C.I.A., John A. McCone, recognized the problem and correctly prophesied which side had the resources to hold out longer:

> We must look with care to our position under a program of slowly ascending tempo of air strikes. With the passage of each day and week we can expect increasing pressure to stop bombing. This will come from various elements of the American public, from the press, the United Nations and world opinion. Therefore, time will run against us in this operation and I think the North Vietnamese are counting on this ... Since the contemplated actions against the North are modest in scale, they will not impose unacceptable damage on it nor will they threaten the DRV's vital interests, hence, they will not present them with a situation with which they cannot live. (quoted in Sharp, 1978, p. 73)

[19] This section is based partly on a student paper of a former NYU undergraduate, Bhashkar Mazumder.

Eventually Rolling Thunder was called off as a result of ever-increasing domestic and international pressure as well as the campaign's failure to stem infiltration from the North, which actually increased significantly (Lewy, 1978, p. 391).

In May 1972, after more than three years of frustration in trying to work out a settlement of the war at peace talks in Paris, President Richard Nixon decided to reassess his options. He could choose between continuing his present course of secret air attacks and public "protective reaction strikes" – at more or less the level of the limited bombing campaign (L) of Lyndon Johnson – or an all-out campaign ($\bar{L}$).

North Vietnam faced a choice of making concessions in response to Nixon's peace plan (C), which would prevent or at least delay the uniting of North and South Vietnam – the North's ultimate goal – or not conceding and, for all practical purposes, demanding U.S. surrender ($\bar{C}$). The preferences of the players for the four possible states, moving clockwise from the upper left-hand state in the "real game" (game 50, which was shown earlier as a class 2 Magnanimity Game in figure 3.5) of figure 4.7, are as follows:

I **Compromise:** (4,3). "Peace with honor" is achieved by the United States, its best state, because "winning" is no longer seen as feasible; also, a negotiated settlement would end domestic and international criticism of the war. Although North Vietnam would not win outright by accepting a negotiated settlement, it would not be devastated by an all-out bombing campaign; moreover, conquering the South would probably be achievable eventually, making this the North's next-best state.

II **Deadlock:** (2,4). The United States remains stymied at the negotiation table, its next-worst state, since the limited bombing campaign, because it permits the North to hold out indefinitely, does not engender serious negotiations. As a consequence, the United States is forced to withdraw without a settlement, which facilitates the quick and forceful unification of Vietnam, giving the North its best state.

III **Destructive Stalemate:** (1,1). This is the worst state for both sides: the United States does not achieve its aim of a settlement and risks a complete collapse of domestic political support because of the extremely adverse public reaction to the massive destruction; the North suffers greatly, with Hanoi and Haiphong leveled and the North's military forces and industrial capacity tremendously damaged.

IV **Limited US Military Victory:** (3,2). This is the next-best state for the United States, because there is a settlement induced by the all-out bombing. But the bombing is seen as unnecessary, even genocidal, if

Real game (game 50)

		North Vietnam			
		Make concessions (C)		Not make concessions ($\bar{C}$)	
	Limit bombing (L)	I Compromise (4,3)^{r*} [4,3]	→	II Deadlock (2,4) [4,3]	←Dominant strategy
United States		↑		↓	
	Not limit bombing ($\bar{L}$)	IV U.S. Military victory (3,2)^{c} [4,3]/[2,4]	←	III Destructive stalemate (1,1) [2,4]/[4,3]	

Game as misperceived by North Vietnam (game 37)

		North Vietnam			
		Make concessions (C)		Not make concessions ($\bar{C}$)	
	Limit bombing (L)	I Compromise (4,3)^{r} [4,3]	→	II Deadlock (3,4)^{c} [3,4]	←Dominant strategy
United States		↑		↓	
	Not limit bombing ($\bar{L}$)	IV U.S. Military victory (2,2) [3,4]	←	III Destructive stalemate (1,1) [3,4]	

Key: (x,y) = (payoff to United States, payoff to North Vietnam)
$[x,y]$ = [payoff to US, payoff to NV] in anticipation game (AG)
4 = best; 3 = next best; 2 = next worst; 1 = worst
c = state Column (North Vietnam) can induce with moving power
r = state Row (United States) can induce with moving power
* = induced state best for both players
Nash equilibria in original games and AGs underscored
NMEs circled
Arrows indicate direction of cycling

Figure 4.7 Vietnam bombing campaigns (games 50 and 37)

North Vietnam signals its willingness to make concessions. Because the North is forced to acquiesce before a settlement is concluded that would relieve it of its suffering, this is its next-worst state.

Before analyzing this game, which is moderately cyclic game 50 in figure 4.4 (the two strategies of the column player are interchanged in figure 4.7), it is worth mentioning some changes that had occurred in the four years that had transpired since the conclusion of Rolling Thunder in 1968. First, the fear on the part of the United States that the Soviet Union and China might intervene (as China had done in the Korean War) was gone; indeed, these countries were now willing to try to persuade North Vietnam to negotiate a settlement. Second, North Vietnam's 1972 Easter invasion of the South had been an utter failure, casting doubt on its military threat to the South.

On the other hand, the United States was very anxious to end the war quickly and no longer insisted on an independent "noncommunist" South; "peace with honor" was sufficient. Without U.S. insistence on a noncommunist South, the North could now gain more from negotiations than previously; and the United States could disentangle itself from an increasingly unpopular war. This is why Compromise is viewed as better for both sides, including the United States, than U.S. military victory.

On May 8, 1972, Nixon announced that its Linebacker bombing campaign had begun (later to be called "Linebacker I"). Although it was not totally unrestricted, it was intended to demonstrate to the North a new and far more destructive level of aerial bombardment if the North continued to be recalcitrant in the peace talks. As recorded by the White House taping system, Nixon said, "The bastards have never been bombed like they're going to be this time" (Palmer, 1978, p. 252).

The destruction, which was accompanied by the mining of Hanoi and Haiphong harbors, was indeed great. In October the North began seriously to negotiate a peace treaty in Paris. Nixon, as a show of good faith, ordered the bombing attacks discontinued on October 23.

But the subsequent negotiations bogged down, largely because of the strong objections of South Vietnamese president Nguyen Van Thieu to an agreement tentatively worked out by Le Duc Tho and Henry Kissinger, who announced on October 26 that "peace was at hand." Although "infuriated by Thieu's intransigence" (Herring, 1979, p. 247), Nixon – after further negotiations with both Thieu and Tho proved fruitless – ordered Linebacker II, making

> absolutely clear to the military his determination to inflict maximum damage on North Vietnam. "I don't want any more of this crap about the fact that we couldn't hit this target or that one," he lectured Admiral Thomas Moorer, Chairman of the Joint Chiefs of Staff. "This is your chance to use military power to win this war, and if you don't, I'll consider you responsible." (Herring, p. 248)

The bombing commenced on December 18, and for the last twelve days of 1972, the North was continually bombarded, around the clock, with almost no restrictions in effect. The effect of this "Christmas bombing" was quickly to produce a peace agreement, which was signed on January 23, 1973, essentially ending U.S. involvement in the war.

To model Linebacker I and Linebacker II, I assume the game originates at Deadlock in the real game of figure 4.7, just before the start of Linebacker I in May 1972. Although the United States had not publicly indicated a limited bombing campaign was under way at this time, there had been secret air attacks as well as a strong air response to the Easter invasion, which I interpret as *L*.

Because the NME from this state is (4,3), induced by the North's moving "magnanimously" from (2,4) to (4,3) (section 3.4),[20] both Linebacker campaigns should not, if rules 5 and 6 were governing, have been necessary. North Vietnam should have switched from $\bar{C}$ to *C*, at which point the United States would have stuck with its choice of *L*.

But the North showed no such magnanimity, primarily because the game was perceived differently from the way I have depicted it (more on this later). Additionally, I believe, the players were not playing according to rules 5 and 6 but instead rules 5′ and 6′.

When the United States commenced Linebacker I, it escalated to Destructive Stalemate. In my view, it was trying to demonstrate its ability to continue cycling – in the clockwise direction indicated in the real game of figure 4.7 – longer than North Vietnam. Although the war had ground on for more than seven years, there was still not the "light at the end of the tunnel" that Kissinger had foreseen earlier.

When North Vietnam responded by choosing its strategy of *C* in

[20] If North Vietnam had not moved, the outcome that the United States could have induced is (3,2), which is worse for both players. Although it does not relate directly to the case being discussed here, there is a subtle issue in this game worth raising at this point. The NME from (1,1) when Column moves first is (3,2) by backward induction, but I assume that the players would agree to stop at the Pareto-superior (4,3), so this state becomes the NME, rather than (3,2), when Column moves first from (1,1), as shown in the anticipation game (AG) in figure 4.7 and in the Appendix. Game 50, it turns out, is the only 2×2 game in which backward induction from a Pareto-inferior state [i.e., (1,1)] gives a Pareto-inferior outcome [(3,2)], except for Prisoners' Dilemma (game 32), in which backward induction from (2,2) gives (2,2). But in Prisoners' Dilemma, neither player would move from the Pareto-inferior (2,2), whereas in game 50 both players have an incentive to move from the Pareto-inferior (1,1), though each would prefer that the other do so first. However, because the outcome that Column can induce by moving first from (1,1) is also Pareto-inferior [(3,2)], it would have an interest in irrevocably committing itself – should Row move past its initial blockage at (3,2) to (4,3) – to stop at (4,3) in the move–countermove process, even though Column can do better by itself moving on from (4,3) to (2,4). In other words, because both players have a common interest in implementing (4,3) rather than (3,2) – assuming that Column moves first from (1,1) – (4,3) would presumably be agreed to by both players, given that they can communicate and commit themselves to a course of action, when Column moves first from (1,1). On the other hand, because an "irrevocable commitment" entails a binding agreement that noncooperative game theory generally does not admit, an argument can also be made that (3,2) should be considered the NME, because Column would have no reason to abide by an agreement once the process reaches (4,3). I do not see an airtight case being made for either (3,2) or (4,3) as *the* NME from (1,1) when Column moves first, which nicely illustrates the nuances that TOM surfaces that the rules of standard game theory keep well submerged.

October 1972, moving to U.S Military Victory at (3,2), the United States in turn responded by choosing *L*, moving the game to Compromise at (4,3). Not only can the United States induce (4,3) with moving power, but North Vietnam also prefers (4,3) to (3,2). The fact that North Vietnam can induce (3,2) if it has moving power, but would not want to do so, makes moving power irrelevant in game 50.

But after Thieu balked, North Vietnam refused to compromise further, seeming not to understand that the possessor of moving power in this game is irrelevant. This moved the game back to the original state of Deadlock (before Linebacker I), completing the cycle.[21]

North Vietnam acted in this manner, I believe, because it misperceived the game being played. It thought that the game was actually game 37 (shown as the misperceived game in figure 4.7), in which moving power is effective (see figure 4.4, in which game 37's two column strategies are interchanged): with it, North Vietnam can induce its best state of Deadlock, which is not only an NME but the unique Nash equilibrium in the anticipation game (AG).

Game 37 occurs when North Vietnam thinks that the United States prefers Deadlock (3) to U.S. Military Victory (2) – rather than vice-versa (as in game 50) – because it believes that the United States would suffer inordinate cost in implementing the latter state. In fact, however, Nixon's later behavior indicated just the opposite, as North Vietnam ruefully was to learn when the Christmas bombing commenced.

Linebacker II was, in effect, Nixon's attempt to demonstrate U.S. moving power once and for all. The Christmas bombing was devastating; it evidently conveyed to the North the message that Nixon was still capable of continuing the cycling, this time by escalating to an unprecedented level of bombing and exerting still greater pressure:

> Such widespread devastation could not be endured. The Politburo got the message. Hanoi quickly agreed to terms, and the bombing stopped abruptly ... Agreement on a cease-fire for all of Vietnam was reached rapidly upon the resumption of the Paris negotiations in January of 1973. (Palmer, 1978, p. 259)

The agreement, however, did not contain "any significant concession from the North" (Papp, 1981, p. 144), differing only on minor points from that negotiated in October (Isaacs, 1983, pp. 61–2).[22] But it had enough face-saving features to enable the United States to claim that its bombing strategy had succeeded.

[21] This cycling echoes the earlier cycling under the Rolling Thunder regime (Thompson, 1980), when bombing and other forms of escalation (e.g., the North's Tet offensive in 1968) were used by both sides to try to maneuver into positions of advantage.

[22] Ball (1993, p. 35) puts it more bluntly: "All that bombing achieved was some very minor changes in an agreement loaded (inevitably, in my opinion) in Hanoi's favor."

The relative effectiveness of the Linebacker operations was substantially due to the fact that the United States had perfected "smart bombs" that could destroy previously indestructible targets, such as the Thanh Hoa bridge, a vital bridge in the North that had been repeatedly attacked to no avail until it was felled with several laser-guided bombs on May 13, 1972. Despite the harsh criticism that Nixon faced for the renewed bombing in both Linebacker operations, he was able to make continuation of the game extremely costly for North Vietnam.[23]

Because North Vietnam misperceived U.S. preferences and, in addition, thought it had the stamina to hold out longer, it forced the United States to prove otherwise, which it did, but with practically no import on the final settlement. Perhaps the greatest tragedy is that the North apparently did not harbor its game 37 view of U.S. preferences at the start of negotiations in 1969, when it offered to settle the war on much the same terms that it accepted in 1973 (Zagare, 1977, p. 683).[24]

By late 1972, however, North Vietnam had come to believe that it not only held the upper hand but also that the United States had lost its will to continue the grinding conflict. By forcing the United States to choose between U.S. Victory and Deadlock – the two states where it has the next move – it believed it could use its moving power in game 37 to induce the latter outcome.

But Nixon's preferences were not those in game 37. Moreover, even if North Vietnam possessed moving power in game 50, it would have been irrelevant in inducing Deadlock. Playing the "wrong" game, the North made what appeared to it as a rational attempt to hold out.

But then came the colossal destruction of the Christmas bombing. Forced to alter its view of the United States as a player bereft of moving power – and, relatedly, that it may have underestimated the U.S. preference for Deadlock (it was 3 rather than 2) – it offered some minor concessions. The United States, which had lost as many as 34 B-52s in the Christmas bombing and faced withering criticism from around the world, was in no position to extract more and so quickly settled (Porter, 1975, pp. 161–5).

[23] In the case of the Christmas bombing, Nixon was accused of being a "madman" and waging a "war by tantrum" (Herring, 1979, p. 248). The "madman" image appears to have been calculated: "President Nixon told one columnist that he ordered the [Christmas] bombings because 'the Russians and Chinese might think they were dealing with a madman and so had better force North Vietnam into a settlement before the world was consumed by a larger war'" (Amter, 1979, pp. 288–9).

[24] In testimony before the Senate Select Committee on P.O.W.–M.I.A. Affairs, Henry Kissinger vehemently denied the contention of Senator John F. Kerry that the terms for a settlement in 1969 were "extraordinarily close" to those in 1973 and, by implication, that the United States was the obstructive party in 1969 (*New York Times*, September 23, 1992, p. A18).

I presume that North Vietnam, because of the Christmas bombing, would have acceded to the Pareto-superior Compromise had it realized that the real game was 50, in which moving power is irrelevant. Note in this game that there are two Pareto-inferior states, (1,1) and (3,2), through which the players must pass as they cycle, as is also true of game 37.[25]

It is also worth noting that (2,4) is the unique Nash equilibrium in game 50, and (3,4) is the unique Nash equilibrium in game 37, which are both associated with the U.S.'s dominant strategy of *L*. Deadlock, then, is the (erroneous) prediction of classical game theory, with or without misperception. By comparison, TOM predicts Compromise as the outcome in both games 37 and 50 when the United States possesses moving power, but only in game 37 is the demonstration of moving power by the United States necessary to induce this state.[26] Because of North Vietnam's misperception, this demonstration occurred, resulting in physical cycling to determine which side, indeed, had greater endurance in the face of massive destruction.[27]

4.6 The effects of reputation

In this chapter I changed rationality rules 5 and 6 to 5′ and 6′ to allow for cycling back to the initial state, with possible repetitions around the matrix. Before inquiring when players would be likely to subscribe to one set of rules over another, let me review some of the consequences of the old rules.

25 Two is the maximum number of such states in a cyclic game; 22 of the 36 cyclic games have two Pareto-inferior states, whereas 12 games have one Pareto-inferior state and 2 games (44 and 25, the two cyclic games of total conflict) have no Pareto-inferior states. There are no cyclic games with three Pareto-inferior states, because such games must contain a (4,4) state, which causes blockage in both directions and thereby precludes cycling (21 of the 78 2×2 games contain such a state). Whether a cyclic game contains zero, one, or two Pareto-inferior states might be considered as an indicator of its simultaneous destructiveness (i.e., to both players at the same time). The greater this destructiveness, the more dramatically players will swing between states of relative cooperation and relative conflict, which makes staying in cooperative states more attractive and cycling, therefore, less likely to continue.

26 In a game with two NMEs, moving power may be thought of as a way for the player that possesses it to induce the choice of the one it prefers, regardless of the initial state, by favoring a "move" over a "stay" at a critical juncture. In effect, the player with moving power rules out "stay" and thus destroys the less-preferred NME, even though (as in both games 50 and 37) this NME may be the Nash equilibrium in both the original game and its AG.

27 Cycling had really begun with Rolling Thunder under President Johnson, of which I have not given a detailed account because this was only the first phase; by 1968, neither side had succeeded in demonstrating that it had moving power. Even after the Christmas bombing in 1972, the apparent supremacy of the United States is deceptive, because North Vietnam ultimately prevailed by successfully invading South Vietnam in 1975 after the withdrawal of U.S. forces.

In chapter 1 I illustrated, as the subtitle proclaims, that "the starting point matters," given there is more than one NME and that players terminate their moves before completing a cycle. But if players can make prior strategy choices in an anticipation game (AG), which gives the NMEs into which each state goes, one NME may stand out. As I showed in chapter 2, one or both players may have dominant–strategy choices associated with it, making it the unique Nash equilibrium in the AG.

In six 2×2 strict ordinal games (42–47) and their AGs, however, neither player has a dominant strategy. Consequently, there is no Nash equilibrium in pure strategies in these games or their AGs, whether the game is one of total conflict or partial conflict. In these games, players' anticipations are therefore of no help in singling out an NME, even if the rules permit them to choose strategies that determine an initial state.

In chapter 3 I showed that not only the starting point matters but also that being magnanimous – by giving up something and, in the process, benefiting the other player – may be rational, starting at Status Quo in some specific games (i.e., those in class 2) of the Magnanimity Game (MG). This result depended on a refinement in the backward-induction argument, based on the two-sidedness convention (TSC), whereby the players take account of the consequences of each other's moves before deciding whether or not to move themselves.

The empirical and other examples I have used to illustrate these consequences of TOM suggest that rules 5 and 6 work well to explain outcomes in many situations. Nonetheless, one observes on occasion essentially the same game being replayed, sometimes with tragic consequences, as in the five Egyptian–Israeli wars and in the repeated bombing campaigns by the United States, and ground offensives by North Vietnam, in the Vietnam war. What rational calculations do the players make in these conflicts to justify the costs of dragging them out for so long?

A contest to determine which, if either, player has moving power provides one explanation. Cyclic games, wherein there is no insuperable barrier to continued movement because no player is ever best off when it chooses next,[28] almost always enable the player with moving power to induce its preferred NME (if there are two in the game), whatever the initial state. This was true in the Revelation Game (RG), although it is reasonable to suppose in this game that it is *SB* who possesses this power. Nevertheless, I showed that if *P* is the player who possesses moving power, it can also induce its best outcome in RG.

[28] But there may be varying degrees of friction in moving from state to state, as I showed in my three-way classification of cyclic games, based on their having zero, one, or two impediments to cycling.

In both the Egyptian–Israeli conflict and the Vietnam war, it seems that the "game" for a long time was over who had moving power. In the end, the Egyptians apparently succeeded in convincing the militarily superior Israelis that they (i.e., the Egyptians) could absorb still another defeat, which would only hurt both sides. On the other hand, both sides would be better off settling their differences rather than going through the painful exercise of another war.

Although three of the four class 1 MG games seemed good candidates to model the protracted Egyptian–Israeli conflict, Magnanimity is not an NME in any of them. This state, however, becomes the rational outcome if the victor accedes to the defeated party's moving power.

Just as it took 31 years for Egypt to establish its claim to this power over Israel, it took eight years for the North Vietnamese to recognize that, especially after the development of smart bombs by the United States, their ability to hold out against the United States had become tenuous. There is a special irony in the latter case, because it was the North Vietnamese for most of the Vietnam war who looked like they would be able to outlast the Americans, for the cogent reasons expressed by John A. McCone (section 4.5) and their ability to launch major ground offensives. And, of course, they did ultimately prevail when they successfully invaded South Vietnam in 1975.

Not all players take years to establish their reputations and have them acknowledged by their opponents. Only three days transpired between the dropping of the first and the second atomic bombs by the United States against Japan in August 1945 before the Japanese signaled their willingness to surrender unconditionally. Regrettably, the Japanese had to suffer grievously before the United States threat was seen as credible.

An already established reputation may substitute for one player's having to make physical moves, sometimes again and again, to make perspicuous its moving power. In the case of the Vietnam war, misinformation about the relative capabilities of each side, especially each's ability to endure continued punishment, was rampant.

If there had been better information about the "real game," in which moving power is irrelevant, there would presumably not have been a battle royal over which player possessed it. There was physical cycling precisely because North Vietnam not only had incomplete information about U.S. preferences but, worse, thought they were different from what they were. If there had not been this misperception, there is evidence that the compromise outcome might have been achieved much earlier and at much less cost to the players.

In Shakespeare's *Othello*, Cassio recognized all too well the calamity of losing his reputation:

Reputation, reputation, reputation! O, I have lost my reputation! I have lost the immortal part of myself, and what remains is bestial.

Undoubtedly, it is harder to regain a reputation, once lost or even tarnished, than to build one up in the first place.

But once a player's reputation is established (or reestablished, as Cassio's eventually was) and acknowledged by an opponent, it may no longer be necessary for players physically to cycle in order to "prove" themselves. Mental moves will then suffice, and a player with recognized moving power may then get its way without suffering the costs of actually cycling.

I shall pursue this idea further in chapter 5, where I analyze the effectiveness of "threat power" in repeated play. I also analyze "order power" in games with indeterminate states, wherein moving power, with only one exception, does not resolve the question of which player can induce a preferred NME from these states.

5 Order and threat power: eliminating indeterminacy and communicating intentions

5.1 Introduction

Power is probably the most suggestive concept in the vocabulary of political scientists. It is also one of the most intractable, bristling with apparently different meanings and implications.

Because power is not a unidimensional concept, it is appropriate to try to incorporate other dimensions – besides the kind of stamina that moving power epitomizes – in different formal definitions. Accordingly, I introduce in this chapter two new kinds of power, "order power" and "threat power," which account for outcomes in games in which the asymmetry in the capabilities of two players is not based on the ability of one to continue moving indefinitely in a cyclic game.

Order power assumes that the player who possesses it can dictate the order of moves from a Pareto-inferior initial state and from which, consequently, both players desire to move. Threat power assumes that one player can threaten the other with the possibility of a Pareto-inferior state – without necessarily moving there – by communicating its intentions in advance.

These concepts will be used to elucidate the flow of moves and exercise of power in two crises, the Cuban missile crisis of 1962 and the Polish crisis of 1980–1, when the independent trade-union movement, Solidarity, temporarily gained the upper hand only to be suppressed later by the Communist party and state. Both crises were resolved without the protagonists resorting to war but instead, as I shall argue, through other demonstrations of power.

What kind of power was decisive in a case, however, and even what game was actually played, are not always evident. In fact, an important lesson of the empirical analysis is that there may not be a single "correct" view of a situation. I shall suggest an alternative view of the Cuban missile crisis from that usually offered and also revisit the Samson and Delilah conflict described in section 1.5.

I start with order power. I introduced the notion of an indeterminate

C (Samson)

R (Delilah)	$\bar{T}$		T	
$\bar{N}$	(2,4) [3,3]	←	(4,2) [4,2]	← Dominant strategy
	↓		↑	
N	(1,1) [2,4]/[3,3]	→	(3,3) [2,4]	

Key: (x,y) = (payoff to R, payoff to C)
$[x,y]$ = [payoff to R, payoff to C] in anticipation game (AG)
4 = best; 3 = next best; 2 = next worst; 1 = worst
N = nag; $\bar{N}$ = don't nag
T = tell secret; $\bar{T}$ = don't tell secret
Nash equilibria in original game and AG underscored
NMEs circled
Arrows indicate direction of cycling

Figure 5.1 Game 56 and its anticipation game (AG)

state in section 1.4, showing that if the initial state in game 56 is (1,1), it will go into (2,4) if R moves first and (3,3) if C moves first. These two bracketed NMEs are shown below (1,1) in game 56's anticipation game (AG) in figure 5.1, indicating the indeterminacy of starting at (1,1). By contrast, there is a single bracketed NME below the other three states of game 56, indicating the determinacy of starting in these other states.[1]

Recall that an AG assumes that rule 5 is governing, so the players are assumed not to complete a cycle in the original game. Thus, if the players start at (1,1) in game 56, each will try to hold out longer – hoping that the other player will be $P1$ and, consequently, move first from (1,1) – because each does better, as $P2$, going second. Specifically, if C is $P2$, it receives its best payoff when R moves first to (2,4), the dominant-strategy Nash equilibrium. On the other hand, if R is $P2$, it receives its next-best payoff when C moves first to (3,3).

In the latter case, R would prefer (4,2) to (3,3), but this outcome is not obtainable from (1,1); (4,2) is the NME only if the initial state is (4,2). Starting at (1,1), the best that R can do is to try to hold out longer there, hoping that C makes the first move to (3,3), which is still better for it than (2,4) should it be forced to move first from (1,1).

The conflict between C and R over which player will move first from (1,1) is resolved if one player possesses *order power*. This is the ability of a

[1] This depiction of game 56 was given in figure 2.3, but I now include the arrows showing that this game cycles in a counterclockwise direction.

player to dictate the order of moves in which the player would depart from an indeterminate initial state to ensure a preferred outcome for itself.[2]

Sometimes, as I shall illustrate later, a player can do better by being the first rather than the second to move from an initial state. Whether a player does better being *P*1 or *P*2, however, I assume that it can make the choice if it has order power.

Order power is applicable only to games with indeterminate states. Because a player in an indeterminate state always benefits from its possession of order power, this power is invariably *effective*: it helps the player who has it, ensuring it of a better outcome than if the other player possessed it.

Order power differs from both "staying power" (Brams, 1983; Brams and Hessel, 1983; Kilgour, De, and Hipel, 1987) and "holding power" (Kilgour and Zagare, 1987) in games that are played according to the rules of TOM. Staying power presumes that the possessor always moves second, whereas holding power presumes that the possessor moves first but can hold or pass, which allows for the possibility of backtracking by the possessor.[3]

Order power is more general than these other concepts in that it does not specify the order of play: at the initial state, the possessor can choose to move either first or second. One consequence of this greater freedom is that its possessor never suffers as the power holder, whereas the possessor of both staying power and holding power does worse in certain games than if the other player were the possessor of this power.

In my opinion, such reversals cast doubt on the usefulness of staying and holding power as concepts that explain why players, because of their special prerogatives, prevail in situations of conflict. If a prerogative may be a liability, degrading the ability of a player to induce the outcomes it prefers, it seems contradictory to call it "power." To be sure, the four games in which moving power is ineffective (section 4.3) also fit this description.

2 Thus, order power is applicable only at the start of play, when a player may or may not move from the initial state. This situation is described by rule 1, after which I assume rules 2–6 apply once order power resolves the indeterminacy at this state.

3 Admittedly, players may make mistakes and decide to backtrack. While backtracking is not permitted by the rules of TOM in games of complete information – wherein players can think ahead and thereby avoid mistakes – TOM does allow for incomplete information by relaxing the assumption of complete information assumed in rules 6 and 6′. Just as incomplete information may cause players to make mistakes, it also may lead them, once they realize their mistakes, to reassess their positions and decide to backtrack. In this sense, TOM can be modified to allow for backtracking in games of incomplete information, which I shall analyze in chapters 6 and 7. Specifically, incomplete information in

In these games, however, there is no rule that prescribes that a player with moving power must force the other player to stop. Rather, it can choose not to exercise this power by stopping itself at a state that it prefers to the (worse) state it can induce with moving power. In this kind of situation, which even ostensibly powerful actors, like a superpower, face on occasion, one's "power" derives from knowing when *not* to exercise it. TOM, in my opinion, clarifies when such situations arise.

The possession of order power, as already noted, always benefits a player when play starts at an indeterminate state. By comparison, threat power can never hurt a player, but it, like moving power, is by no means effective in all games, as I shall show later.

5.2 The interplay of different kinds of power

To illustrate the concept of order power, consider again game 56, which I analyzed at length in chapter 1. In section 1.5 I used this game to illustrate the conflict between Samson and Delilah, wherein (1,1) represented the state, "Delilah frustrated, Samson harassed," when Delilah nagged Samson (N) and Samson refused to tell the secret of his strength ($\bar{T}$) (figure 1.2).

Suppose in this game that Delilah, as R in figure 5.1, has order power when the players find themselves stuck at (1,1). By continuing to nag, she can induce Samson, as C, to depart first from (1,1), telling the secret of his strength, which yields the outcome (3,3).

Now assume that Delilah has moving power in game 56, and that rule 5′, which permits cycling, is governing. Then she can present Samson with a choice between (4,2) and (1,1), where he, as C, has the next move when the game cycles counterclockwise, as shown in figure 5.1. With moving rather than order power, Delilah can actually induce her best state of (4,2), which is the outcome, "Delilah happy, Samson forthcoming," when Samson tells his secret and she quits nagging.

Thus, given rules 1–4, order power (rules 5 and 6 operative) and moving power (rules 5′ and 6′ operative) make different predictions in game 56 when R (Delilah) possesses one or the other kind of power. Obviously, she would prefer to have moving power, obtaining her best outcome of (4,2) when C (Samson) tells his secret without her having to continue to nag. But remember that she obtains this outcome only by having been a nag earlier, when the game cycled, after which Samson finally succumbed to her wiles.

The fact that this game cycled three times in the story lends credibility

the Magnanimity Game (MG) is discussed in section 6.3, and backtracking in MG is discussed in section 7.1.

to the moving-power interpretation. Nevertheless, I think it equally persuasive to say that the players quickly became stuck at (1,1). Once there, Samson eventually caved in, letting the truth spill out. By continuing to nag and wear down his resistance, Delilah, according to the order-power interpretation, was finally able to ensnare him at (3,3).

I offered neither an order-power [giving (3,3)] nor a moving-power [giving (4,2)] interpretation of the outcome of this conflict in section 1.5. Instead, I suggested that, starting at (2,4) – before Delilah begins nagging – the players would move to (3,3) if, according to rules 5 and 6, cycling is prohibited.

But this is the same outcome as if Delilah had order power and play commenced at (1,1). Thus, there are competing explanations for how (3,3) might have arisen. On the other hand, if one believes that cycling occurred and rules 5′ and 6′ are therefore applicable, then Delilah, with moving power, can ensure (4,2) from any initial state.[4]

I offer these different interpretations of the Samson-and-Delilah story, based on TOM, not to single out one as indisputably "correct." In my opinion, all have merit. To summarize,

(i) from initial state (2,4), the process goes into (3,3), based on rules 5 and 6, with no assumption about a more powerful player;
(ii) from (1,1) Delilah, with order power, can induce (3,3), based on rules 5 and 6;
(iii) independent of the initial state, Delilah, with moving power, can induce (4,2), based on rules 5′ and 6′.

Fitting an interpretation to a story is, to a degree, a matter of taste. I leave to the reader the choice of which of the three interpretations he or she thinks most naturally explains the outcome of the biblical story.

My purpose in introducing order power into the analysis is not simply to provide a new gloss on game 56, especially when other concepts may provide an equally good explanation of the outcome in the Samson-and-Delilah story. If this were the case, then it would be preferable, for reasons of parsimony, to dispense with order power.

In fact, however, order power resolves indeterminacies in several games in which moving power is irrelevant (games 49 and 50) or undefined because games are noncyclic (games 51–55, 57). Indeed, game 56 is the only 2×2 strict ordinal game in which both moving and order power are effective. In the other games with indeterminate states, order power offers a fundamentally different slant on the choice of different outcomes, at least when the initial state is indeterminate, so it fills a gap when a

4 By contrast, if Samson possesses moving power, he can ensure (3,3), so moving power, like order power from (1,1), is effective in game 56.

determinate prediction is desired and there is an asymmetry in the players' capabilities.[5]

The best known of the indeterminate games, Chicken (game 57), is symmetric, which makes it noncyclic, so moving power is undefined in it. Thus, this concept of power cannot offer any insight into the selection of one outcome over another, starting at an indeterminate state, in this game. The problem of outcome selection is compounded by the fact that Chicken has three NMEs, which makes it the only game, besides game 56, to have so many.

In section 5.4 I shall use Chicken, and suggest another 2×2 game (game 30), as possible models of the 1962 Cuban missile crisis. Before doing so, however, I shall recount the stories of both Chicken and Prisoners' Dilemma (game 32) in section 5.3 and describe the strategic features of these renowned games that have made them so fascinating to theorists.

Chicken and Prisoners' Dilemma are often compared in the game-theoretic literature, which I shall also do here. I begin by interpreting them according to the standard theory and then according to TOM. Differences between the predictions of these theories, based on the different rules, are striking.

According to TOM, order power is not applicable to Prisoners' Dilemma because it has no indeterminate states. Neither is moving power, because Prisoners' Dilemma, like Chicken, is noncyclic. However, a third kind of power, "threat power," which posits a threatener who can communicate its intentions, is.

Threat power assumes repeated play of a game. If possessed by one player in Prisoners' Dilemma, it works to the advantage of both players in distinguishing the Pareto-superior NME from the Pareto-inferior NME.[6] In this manner, the dilemma in this infamous game is resolved not simply by good will but by the threat of harm if cooperation on the part of the players is not forthcoming.

I shall distinguish two different kinds of threats in section 5.5, "compellent" and "deterrent," whose exercise requires another modification in the rules of play of TOM. I illustrate this distinction with game 22 in section 5.6, which I use to model the Polish crisis of 1980–1, especially its shifting outcomes over time.

This analysis shows how the earlier perspective of TOM – of starting play in a state – can be expanded to include the communication of

[5] When the players' capabilities are more or less equal, then it is the initial state that is the crucial variable if there is more than one NME and no indeterminate state.

[6] Threat power, as well as moving power, also singles out the cooperative (3,3) state in game 30, which is the other game, besides Chicken, that I posit as a possible model of the Cuban missile crisis.

intentions. This communication, if credible, becomes part of the structure of a game and affects future strategy choices when play is assumed to be repeated. I conclude in section 5.7 with some reflections on the role of power, in its different varieties, on the play of games.

5.3 Prisoners' Dilemma and Chicken

Prisoners' Dilemma (or PD, which is game 32) and Chicken (game 57) are two of the 12 games subsumed by the generic Magnanimity Game (MG), all of which are listed in figure 3.5. Because of their symmetry, these two games are noncyclic games, so moving power is not defined in them.

Although PD and Chicken are the only specific MG games that are symmetric, they were not earmarked for special analysis in section 3.4. Whereas PD is a class 2 game and Chicken is a class 3 game, it is in the interest of the victor in both games to be magnanimous.

In the figure 5.2 representation, the players in both of these games are assumed to be able to choose between the strategies of cooperation (C) and noncooperation ($\bar{C}$). The short-hand verbal descriptions given in figure 5.2 for each state are intended to convey their qualitative nature, based on the players' rankings.

In both games, each player obtains its next-best payoff of 3 by choosing C if the other player also does (Compromise), but both have an incentive to defect from this state to obtain their best payoffs of 4 by choosing $\bar{C}$ when the other player chooses C. Yet if both choose $\bar{C}$, they bring upon themselves their next-worst state (Conflict) in PD, and their worst state (Disaster) in Chicken. Clearly, PD and Chicken are games of partial conflict, in which both players can, together, do better at some states (e.g., CC) rather than others ($\bar{C}\bar{C}$).

The dilemma in PD, according to the standard theory, is that both players have dominant strategies of choosing $\bar{C}$: whatever the other player chooses (C or $\bar{C}$), $\bar{C}$ is better; but the choice of $\bar{C}$ by both leads to (2,2), which is Pareto-inferior to (3,3). In addition, (2,2) is a Nash equilibrium, because neither player has an incentive to depart unilaterally from this outcome because it would do worse if it did.

Presumably, rational players would each choose their dominant, or unconditionally best, strategies of $\bar{C}$ in PD, resulting in the Pareto-inferior (2,2) Nash equilibrium, which has been called a *trap* (Cross and Guyer, 1980; Brams, 1985a, chapter 7). Because of its stability, and despite (3,3)s being better for both players, neither player would be motivated to depart from (2,2). Should (3,3) somehow manage to be chosen, however, both players would be tempted to depart from it to try to do still better, rendering it unstable. Put another way, mutual cooper-

Prisoners' Dilemma (game 32)

Row \ Column	C	$\bar{C}$	
C	Compromise (3,3) [3,3]	Column wins (1,4) [3,3]	
$\bar{C}$	Row wins (4,1) [3,3]	Conflict (2,2) [2,2]	← Dominant strategy
		↑ Dominant strategy	

Chicken (game 57)

Row \ Column	C	$\bar{C}$
C	Compromise (3,3) [3,3]	Column advantaged (2,4) [4,2]/[3,3]
$\bar{C}$	Row advantaged (4,2) [3,3]/[2,4]	Disaster (1,1) [2,4]/[4,2]

Key: (x,y) = (payoff to Row, payoff to Column)
[x,y] = [payoff to Row, payoff to Column] in anticipation game (AG)
4 = best; 3 = next best; 2 = next worst; 1 = worst
Nash equilibria in original games and AGs underscored
NMEs circled

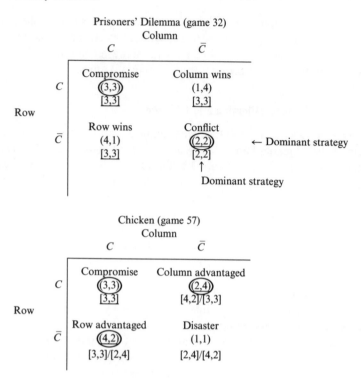

Figure 5.2 Prisoners' Dilemma (game 32) and Chicken (game 57)

ation would seem hard to sustain because each player has an incentive to double-cross its opponent and switch to $\bar{C}$.

The difficulties of cooperation in PD can perhaps be made more vivid by recounting the story that gives Prisoners' Dilemma its name, attributed to A.W. Tucker. Two persons suspected of being partners in a crime are arrested and placed in separate cells so that they cannot communicate with each other. Without a confession from at least one suspect, the district attorney does not have sufficient evidence to convict them of the crime. In an attempt to extract a confession, the district attorney tells each suspect the following consequences of their (joint) actions:

1 If one suspect confesses ($\bar{C}$ in figure 5.2) and the other one does not (C), the one who confesses can go free (gets no sentence) for cooperation

with the state, but the other gets a stiff 10-year sentence – equivalent to (4,1) and (1,4) in figure 5.2.

2 If both suspects confess ($\bar{C}\bar{C}$), both get reduced sentences of 5 years – equivalent to (2,2) in figure 5.2.

3 If both suspects remain silent (CC), both go to prison for 1 year on the lesser charge of carrying a concealed weapon – equivalent to (3,3) in figure 5.2.

What should the suspects do to save their own skins, assuming that neither has any compunction against "squealing" on the other? Observe first that if either suspect confesses, it is advantageous for the other suspect to do likewise to avoid the very worst outcome of ten years in prison. The rub is that confessing and receiving a moderate sentence of five years is not at all appealing, even though neither suspect can assure itself of a better outcome. Moreover, if one's partner does not confess, it is also better to turn state's evidence by confessing and, therefore, to go free.

Thus, both suspects' strategies of confessing ($\bar{C}$) strictly dominate their strategies of not confessing (C), though the choice of the former strategy by both suspects leads to a worse outcome (five years in prison for each) than maintaining silence and getting only one year in prison. But without being able to communicate with one's partner to coordinate a joint strategy – much less make a binding agreement – a suspect could, by not confessing, set himself or herself up for being double-crossed. Indeed, in contemplating the "cooperative" outcome of not squealing (CC), both suspects will be tempted to defect, impelling them to choose $\bar{C}\bar{C}$.

One might think that if PD were played repeatedly, perspicacious players could, in effect, communicate with each other by establishing a pattern of previous choices that would reward the choice of the co-operative strategy. But if the game ends after n rounds, and the final round is therefore definitely known, it clearly does not pay to cooperate on this round since, with no plays to follow, the players are in effect in the same position if they played the game only once. If there is no point in trying to induce a cooperative response on the nth round, however, such behavior on the $(n-1)^{\text{st}}$ round would be to no avail, either, since its effect could extend only to the nth round, when cooperative behavior has already been ruled out. Carrying this reasoning successively backwards, it follows that one should not choose the cooperative strategy in any round of the game.

In section 5.5 I shall explore further this line of reasoning when threats are possible, suggesting how "threat power" can stabilize the cooperative outcome. But next I turn to the problems that players face in Chicken, according to the standard theory.

In this game there are two Nash equilibria in pure strategies, (4,2) and (2,4), both of which are Pareto-optimal – because there are no other states better for both players – and Pareto-superior to (1,1). But each player, in choosing its strategy $\bar{C}$ associated with the Nash equilibrium favorable to itself [(4,2) for Row, (2,4) for Column], risks the selection of the disastrous (1,1) outcome should the other player also choose $\bar{C}$.

The fact that neither player has a dominant strategy in Chicken means that the better strategy choice of each (C or $\bar{C}$) depends on the strategy choice of the other player. This interdependence gives each player an incentive to threaten to choose $\bar{C}$, hoping that the other will concede by choosing C so that the threatener can obtain its preferred Nash equilibrium. As in PD, the compromise (3,3) outcome is unappealing because, should it be chosen, it is not stable.

The story usually told about Chicken is the following: Two drivers race toward each other on a narrow road. Each has the choice between swerving, and definitely avoiding a head-on collision (C), or continuing on the collision course ($\bar{C}$). Each player would

- most prefer that the other player "chicken out" when it does not (4) at $C\bar{C}$ or $\bar{C}C$;
- next most prefer that both chicken out, making the disgrace of doing so milder (3) at CC;
- next least prefer to be the sole player to chicken out (2) at $C\bar{C}$ or $\bar{C}C$; and
- least prefer that neither player chickens out, causing both to die in a head-on collision (1), at $\bar{C}\bar{C}$.

Speaking abstractly, it is hard to say which game, PD or Chicken, poses more obdurate problems for the players. Therefore, I turn next to an application of Chicken – and later a second game – to the Cuban missile crisis, after which I will consider PD further as well as an asymmetric game that models the Polish crisis of 1980–1. I shall argue that order power helps players choose cooperation in Chicken, and threat power helps them escape the dilemma in PD.

5.4 The Cuban missile crisis: moving and order power

Before the breakup of the Soviet Union in 1991 and its demise as a superpower, the most dangerous confrontation between the superpowers ever to occur was the Cuban missile crisis, which was known as the Caribbean crisis in the Soviet Union. It was precipitated by a Soviet attempt in October 1962 to install in Cuba medium-range and intermediate-range nuclear-armed ballistic missiles capable of hitting a large portion of the United States.[7] How the superpowers – or, more accurately, their leaders – managed this crisis has been described in great detail, but this scrutiny has involved little in the way of formal and systematic strategic analysis.

After the presence of such missiles was confirmed on October 14, the Central Intelligence Agency estimated that they would be operational in about ten days. A so-called Executive Committee (ExCom) of high-level officials was convened to decide on a course of action for the United States, and ExCom met in secret for six days. Several alternatives strategies were considered, which were eventually narrowed to the two that I shall discuss.

The most common conception of this crisis is that the two superpowers were on a collision course. Chicken, at first blush, would seem an appropriate model of this conflict. As applied to the Cuban missile crisis, with the United States and the Soviet Union the two players, the alternative courses of action, and a ranking of the players' states in terms of the game of Chicken, are shown in figure 5.3.[8]

The goal of the United States was immediate removal of the Soviet missiles, and United States policy makers seriously considered two strategies to achieve this end:

1 A naval blockade (*B*), or "quarantine" as it was euphemistically called, to prevent shipment of further missiles, possibly followed by stronger action to induce the Soviet Union to withdraw those missiles already installed.
2 A "surgical" air strike (*A*) to wipe out the missiles already installed, insofar as possible, perhaps followed by an invasion of the island.

The alternatives open to Soviet policy makers were:

1 Withdrawal (*W*) of their missiles.
2 Maintenance (*M*) of their missiles.

Needless to say, the strategy choices and probable outcomes as presented in figure 5.3 provide only a skeletal picture of the crisis as it developed over a period of thirteen days. Both sides considered more than the two alternatives listed, as well as several variations on each. The Soviets, for example, demanded withdrawal of American missiles from Turkey as a *quid pro quo* for withdrawal of their own missiles from Cuba, a demand publicly ignored by the United States.

7 This section is drawn in part from Brams (1977) with permission; deception possibilities in this crisis are analyzed in section 6.5.

8 I assume that the superpowers can be considered unitary actors, though this is an obvious simplification. It is rectified in part by constructing alternative models that emphasize different features of the crisis, as Allison (1971) has done.

United States (*US*) / Soviet Union (*SU*)	Withdrawal (*W*)	Maintenance (*M*)
Blockade (*B*)	Compromise (3,3) [3,3]	Soviet victory, U.S. defeat (2,4)# [4,2]/[3,3]
Air strike (*A*)	U.S. victory, Soviet defeat (4,2)* [3,3]/[2,4]	Nuclear war (1,1) [2,4]/[4,2]

Key: (x,y) = (payoff to *US*, payoff to *SU*)
[x,y] = [payoff to *US*, payoff to *SU*] in anticipation game (AG)
4 = best; 3 = next best; 2 = next worst; 1 = worst
Nash equilibria in original games and AGs underscored
NMEs circled
* = *US*'s compellent threat state
# = *SU*'s compellent threat state

Figure 5.3 Cuban missile crisis (game 57)

Furthermore, there is no way to verify that the states given in figure 5.3 were the most likely ones, or valued in a manner consistent with the game of Chicken. For example, if the Soviet Union had viewed an air strike on their missiles as jeopardizing their vital national interests, the (4,2) state may very well have ended in nuclear war between the two sides, giving it the same value as (1,1). Still another simplification relates to the assumption that the players chose their actions simultaneously, when in fact a continual exchange of messages, occasionally backed up by actions, occurred over those fateful days in October.

Nevertheless, most observers of this crisis believe that the two superpowers were on a collision course, which is actually the title of one book describing this nuclear confrontation.[9] Most observers also agree that neither side was eager to take any irreversible step, such as one of the drivers in Chicken might do by defiantly ripping off its steering wheel in full view of the other driver, thereby foreclosing the option of swerving.

Although in one sense the United States "won" by getting the Soviets to withdraw their missiles, Premier Nikita Khrushchev at the same time extracted from President John Kennedy a promise not to invade Cuba,

[9] Pachter (1963). Other books on this crisis include Abel (1966); Weintal and Bartlett (1967); Kennedy (1969); Allison (1971); Divine (1971); Chayes (1974); Dinerstein (1976); Detzer (1979); and Brune (1985). Books that take account of revelations from the Soviet side include Blight and Welch (1989), Garthoff (1989), and Thompson (1992); for details on the use of aerial surveillance, see Brugoni (1992).

which seems to indicate that the eventual outcome was a compromise solution of sorts.[10] Moreover, even though the Soviets responded specifically to the blockade and, therefore, did not make the choice of their strategy independently of the American strategy choice, the fact that the United States held out the possibility of escalating the conflict to at least an air strike indicates that the initial blockade decision was not considered final – that is, the United States considered its strategy choices still open after imposing the blockade.

Truly, this was a game of sequential bargaining, in which each side did not make an all-or-nothing choice but considered alternatives should the other side fail to respond in a manner considered appropriate. Representing the most serious breakdown in the deterrence relationship between the superpowers that had persisted from World War II until that time, each side was gingerly feeling its way, step by ominous step.

Before the crisis, the Soviets, fearing an invasion of Cuba by the United States and also the need to bolster their international strategic position, concluded that it was worth the risk of installing the missiles; confronted by a *fait accompli*, the United States, in all likelihood, would be deterred from invading Cuba and would not attempt any other severe reprisals (Garthoff, 1989). Presumably, the Soviets did not reckon the probability of nuclear war to be high in making their calculation,[11] thereby making it rational for them to risk provoking the United States.

Although this thinking may be more or less correct, there are good reasons to believe that U.S. policy makers viewed the game not to be Chicken at all, at least as far as they ranked the possible states. In figure 5.4, I offer an alternative representation of the Cuban missile crisis,[12] retaining the same strategies for both players as given in the Chicken representation (figure 5.3) but assuming a different ranking of states by the United States. The resulting game is 30, whose states may be interpreted as follows:

1 **BW**: The choice of blockade by the United States and withdrawal by the Soviet Union remains the compromise state for both players – (3,3).

10 A release of letters between Kennedy and Khrushchev, however, indicates that Kennedy's promise was conditional on the good behavior of Cuba (Pear, 1992).

11 This probability seems to have been higher than initially thought, according to recent disclosures, which indicate that the Soviets had four times as many troops as U.S. intelligence estimated as well as tactical nuclear weapons that they would have used in the event of a U.S. invasion (Tolchin, 1992). At the height of the crisis, President Kennedy estimated the chances of war to be between one-third and one-half (Sorensen, 1965, p. 705). For a theoretical analysis of the probability of nuclear war, see Avenhaus, Brams, Fichtner, and Kilgour (1989).

12 Still a different 2×2 game is proposed in Snyder and Diesing (1977, pp. 114–16); an "improved metagame analysis" of the crisis is presented in Fraser and Hipel (1982–3), and a game-tree analysis is offered in Brams (1985a, 1985b, 1990).

	Soviet Union (SU) Withdrawal (W)		Soviet Union (SU) Maintenance (M)
Blockade (B)	Compromise (3,3)*# [3,3]	→	Soviet victory, U.S. capitulation (1,4) [3,3]
United States (US)	↑		↓
Air strike (A)	"Dishonorable" U.S. action, Soviets thwarted (2,2) [3,3]	←	"Honorable" U.S. action, Soviets thwarted (4,1) [3,3]

Key: (*x*,*y*) = (payoff to *US*, payoff to *SU*)
[*x*,*y*] = [payoff to *US*, payoff to *SU*] in anticipation game (AG)
4 = best; 3 = next best; 2 = next worst; 1 = worst
NMEs circled
* = state induced by moving power of *US* that is better for both *US* and *SU* than state, (2,2), *SU* can induce
# = US's deterrent threat state and *SU*'s compellent threat state
Arrows indicate direction of cycling

Figure 5.4 Cuban missile crisis (game 30)

2 **BM**: In the face of a U.S. blockade, Soviet maintenance of their missiles leads to a Soviet victory (its best state) and U.S. capitulation (its worst state) – (1,4).
3 **AM**: An air strike that destroys the missiles that the Soviets were maintaining is an "honorable" U.S. action (its best state) and thwarts the Soviets (their worst state) – (4,1).
4 **AW**: An air strike that destroys the missiles that the Soviets were withdrawing is a "dishonorable" U.S. action (its next-worst state) and thwarts the Soviets (their next-worst state) – (2,2).

Even though an air strike thwarts the Soviets in the case of both states (2,2) and (4,1), I interpret (2,2) to be a less damaging state for the Soviet Union. This is because world opinion, it may be surmised, would severely condemn the air strike as a flagrant overreaction – and hence a "dishonorable" action of the United States – if there were clear evidence that the Soviets were in the process of withdrawing their missiles, anyway. On the other hand, given no such evidence, a U.S. air strike, perhaps followed by an invasion, would probably be viewed by U.S. policy makers as a necessary, if not "honorable," action to dislodge the Soviet missiles.

Before analyzing these possibilities, however, I shall offer a brief justification – mainly in the words of the participants – for the alternative representation given by game 30. The principal protagonists, of course, were President Kennedy and Premier Khrushchev, the leaders of the two countries. Their public and private communications over the thirteen days of the crisis indicate that they both understood the dire consequences of precipitous action and shared, in general terms, a common interest in preventing nuclear war. For the purpose of the present analysis, however, what is relevant are their specific preferences for each state.

Did the United States prefer an air strike (and possible invasion) to the blockade (and its eventual removal), given that the Soviets would withdraw their missiles? In responding to a letter from Khrushchev, Kennedy said:

> If you would agree to remove these weapons systems from Cuba ... we, on our part, would agree ... (a) to remove promptly the quarantine measures now in effect and (b) to give assurances against an invasion of Cuba. (Allison, 1971, p. 228)

This statement is consistent with the game 30 representation of the crisis [since (3,3) is preferred to (2,2) by the United States] but not consistent with the Chicken representation [since (4,2) is preferred to (3,3) by the United States].

Did the United States prefer an air strike to the blockade, given that the Soviets would maintain their missiles? According to Robert Kennedy, a close adviser to his brother during the crisis, "If they did not remove those bases, we would remove them" (Kennedy, 1969, p. 170). This statement is consistent with the game 30 representation [since (4,1) is preferred to (1,4) by the United States] but not consistent with the Chicken representation [since (2,4) is preferred to (1,1) by the United States].

Finally, it is well known that several of President Kennedy's advisers felt very reluctant about initiating an attack against Cuba without exhausting less belligerent courses of action that might bring about the removal of the missiles with less risk and greater sensitivity to American ideals and values. As Robert Kennedy put it, an immediate attack would be looked upon as "a Pearl Harbor in reverse, and it would blacken the name of the United States in the pages of history" (Sorensen, 1965, p. 684). This statement is consistent with the United States' ranking *AW* next worst (2) – a "dishonorable" U.S. action in the game 30 representation – rather than best (4) – a U.S. victory in the Chicken representation.

If game 30 (figure 5.4) provides a more realistic representation of the participants' perceptions than does Chicken (figure 5.3), the standard theory offers little in the way of explanation of how the Compromise (3,3) state was achieved and rendered stable in either game. After all, as in

Chicken, this state is not a Nash equilibrium in game 30; but unlike Chicken, no other state in game 30 is a Nash equilibrium in pure strategies.

The instability of states in this game can most easily be seen by examining the cycle of preferences, indicated by the arrows going in a clockwise direction in game 30. Following these arrows shows that this game is strongly cyclic (section 4.2), with one player always having an immediate incentive to depart from every state: the Soviets from (3,3) to (1,4); the United States from (1,4) to (4,1); the Soviets from (4,1) to (2,2); and the United States from (2,2) to (3,3).

Chicken, as noted in section 5.2, is noncyclic, with its Nash equilibria of (4,2) and (2,4) precluding cycling in a clockwise and a counterclockwise direction, respectively. What Chicken and game 30 share, according to the standard theory, is that neither player has a dominant strategy: each player's best strategy depends on the strategy choice of the other player. In Chicken, for example, the United States prefers *B* if the Soviets choose *W*, but *A* if they choose *M*.

How, then, can one explain the choice of (3,3) in either game, given its nonequilibrium status according to the standard theory? The evident answer in game 30, according to TOM, is that (3,3) is the unique NME: wherever play commences, players will move to this state if they are not there in the first place.

But this explanation is perhaps a little too glib, because it assumes that there is a prohibition on cycling, according to rules 5 and 6. As I indicated earlier, however, bargaining between the players went back and forth, suggesting that the players were willing to revisit previous states, especially if it might save them from a nuclear holocaust (more evidence on this point will be given in section 6.5, where I discuss the possible use of deception by Khrushchev in the crisis).

Because the conflict occurred in the Caribbean, within the U.S. "sphere of influence," the United States could bring to bear far greater military capabilities than the Soviet Union. Presuming that these capabilities give the United States moving power in game 30, they enable it to induce the Soviets to stop at either (3,3) or (4,1), where they have the next move (see figure 5.4). Clearly, the Soviets would prefer "compromise" at (3,3) to being "thwarted" at (4,1).

Reinforcing this choice is the fact that moving power is irrelevant in this game. If the Soviets possessed it (unlikely as this may be), they can force the United States to choose between (2,2) and (1,4). But instead of inducing (2,2), it would be in the Soviets interest to accede to (3,3). Thus, if game 30 cycles, as allowed by rules 5′ and 6′, then "compromise" is the only rational choice of the players, whichever has moving power.

In fact, the United States not only had superior military capabilities in this confrontation, but President Kennedy was also determined to show his resolve in this crisis after the 1961 Bay of Pigs debacle in Cuba. Hence, I believe moving power offers a persuasive explanation of the outcome in this game, at least insofar as rules 5′ and 6′, rather than rules 5 and 6, were governing.[13]

Moving power is undefined in Chicken because it is a noncyclic game. In this game, which has three NMEs, it is order power that is crucial in singling out an outcome in all states except (3,3). If the players at the height of the crisis are at (1,1), for example, then the U.S.'s order power induces (4,2), a U.S. victory and Soviet defeat (figure 5.3) by forcing the Soviets to back off from *M* and move to *W*.

In fact the Soviets retreated first, but they did not do so in response to a U.S. air strike, as suggested in the Chicken representation of figure 5.3. Rather, only the blockade was in effect, even at the height of the crisis, so it is reasonable to assume that the game begins at (2,4), with Soviet missiles not yet removed.

In this situation, paradoxically, it is the Soviets' order power in Chicken that enables them to move to (3,3), given that the United States does not first escalate to an air strike at (1,1) to induce a subsequent move by the Soviets to (4,2). Yet it is highly unrealistic to suppose that the Soviets had order power, once their missiles had been detected and the U.S. blockade had been imposed.

Assume neither player has order power in the Chicken representation. Even without such power, as I showed in section 3.4 (figure 3.5), the NME induced in Chicken from (2,4) will always be a state associated with the strategy of "magnanimity," which is *W* in the Cuban missile crisis. And in this game, it is rational for the Soviets to take the first step, moving from (2,4) to (3,3) [equivalently, from (4,2) to (3,3) in figure 3.5].

Knowing that an escalation of the conflict by the United States would have been all but certain had they not retreated first, the Soviets chose to be magnanimous. Indeed, if they had not been, it would have been rational for the United States to have induced (4,2) in Chicken by moving from (2,4) to (1,1), from which the Soviets would then move to (4,2) – unless (1,1) were catastrophic, such as an all-out nuclear war, preventing movement away from it.

The fact that the United States delayed an air strike suggests that it did not want to take the chance of trying to move play "through" (1,1), which

[13] Since a move by the United States from air strike to blockade was probably infeasible, the correspondence between Kennedy and Khrushchev might be interpreted as verbal moves or probings, by which both sides could back off from their threats in a feasible manner before making physical moves (for more on their correspondence, see section 6.5).

might not have been possible if escalation to nuclear war had occurred at this point. Occasionally, then, players might not want to exercise order power if it entails movement that, realistically, might be either risky or infeasible (section 1.6).

In game 30, on the other hand, the moving-power explanation reinforces the choice of the NME of (3,3) in this game. Chicken, by comparison, works best as a model if the Soviets are assumed, starting at (2,4) in figure 5.3, to make the rational choice of "magnanimity" – though not after winning a war – in this specific example of an MG game. At the same time, it was in the interest of the United States to "let" the Soviets be magnanimous so as to be able to bypass (1,1).[14]

The two different representations of the Cuban missile crisis testify to the need to ponder strategic conflicts from different perspectives that take into account the perceptions of players, as was done in the case of the bombing campaigns in Vietnam in section 4.5 (another example of misperception – in the Iran hostage crisis – will be given in section 6.4). In the fashion of *Rashomon* (a Japanese movie that portrays four different versions of a rape), each perspective gives new insights. It is especially instructive to see how sensitive outcomes are to the different reconstructions on which each is based and the relationship of these to the actual outcome.[15] In section 6.5, I will return to the Cuban missile crisis to explore the possibility of deception in it. But next I introduce a new power concept.

5.5 Threat power

The exercise of "threat power" offers an alternative explanation of the outcome in game 30. Informally (this concept will be formalized later), threat power gives a player greater ability to endure a Pareto-inferior state than its opponent, should it have to carry out a threat that hurts both

[14] In effect, letting the Soviets back down by moving to (3,3) is to carry the concept of order power to one higher level. If traversing (1,1) were not so hazardous, it would clearly be in the interest of the United States to depart first from (2,4) to induce (4,2), its best state. [Notice that each player's desire to move first from (2,4) is the opposite of its desire to move second from (1,1).] But because of the danger of nuclear war at (1,1), it was rational for the United States to prescribe a different order of moves, giving the Soviets the opportunity to seek a compromise under the threat of more dire action than the blockade. I shall say more about the use of threats in section 5.5.

[15] A politico-military crisis that erupted into war, and for which different game-theoretic representations have been proposed, is that between Great Britain and Argentina in 1982 over the control of the Falkland/Malvinas Islands. See Sexton and Young (1985), Zagare (1987, pp. 48–9 and 56–62), Thomas (1987), Bennett (1987), and Hipel, Wang, and Fraser (1988). Game-theoretic representations of several other crises are given in Snyder and Diesing (1977).

players; by prevailing, on the other hand, it can induce the choice of a Pareto-superior outcome, which is, of course, preferred by both players.

In game 30 (see figure 5.4), for example, by threatening to choose *A*, which includes the Soviet Union's two worst states (1 and 2), the United States can induce the Soviets to choose *W* when the United States chooses *B*, resulting in (3,3). Even though the Soviets have an incentive to move from (3,3) to (1,4), as indicated by the top horizontal arrow, they would be deterred from doing so by the threat that if they did, the United States would choose its strategy *A* and – by virtue of possessing threat power – stay there. Assuming the Soviets move to their preferred state in the second row, the United States would still inflict upon them (2,2), which is Pareto-inferior to (3,3).

Given that the United States has threat power, then, it is rational for the Soviets to accede to this threat, enabling the United States to implement (3,3). If the Soviets possess threat power in game 30, as I shall show later, the outcome would not change, making who possesses threat power irrelevant in this game. However, though irrelevant, its impact is certainly salutary in allowing the players to avoid (2,2) in a game in which neither player has a dominant strategy, there are no Nash equilibria (in pure strategies), and, consequently, there is no indubitably rational choice according to the standard theory.

Not only does the existence of threat power lead to the choice of (3,3) in game 30, but it does so in other games, including PD (game 32). Unlike Chicken, PD has no indeterminate states, so order power is irrelevant in this game. So is moving power, because PD is noncyclic. And if play starts at (2,2), this state becomes the NME; for (3,3) to be the NME, play would have to commence at one of the other three states.

An escape from (2,2) comes via the exercise of threat power by one player, which is of the same deterrent variety as the United States can use in game 30. Assume Row threatens to choose $\bar{C}$ in PD – giving Column a payoff of either 2 or 1 – unless Column chooses *C* when Row does, giving both players a payoff of 3. Plainly, it is rational for Column to accede to this threat to prevent the selection of (2,2). Thereby the exercise of threat power in PD, based on a policy of tit-for-tat (cooperate if and only if the other player does), induces cooperation. This result, which I shall formalize shortly, echoes other resolutions of PD that assume repeated play without a fixed number of rounds and discounting of future play.[16]

Note that threats offer a resolution of PD very different from the trust that some theorists have argued is necessary, in the absence of a binding and enforceable contract, to induce cooperation in this well-nigh intracta-

16 Axelrod (1984); for criticisms of this work, see Coll and Hirshleifer (1991).

ble game. Indeed, when players begin by choosing strategies in this game without making threats and then have no possibility of making moves and countermoves later, it is simply not rational to be either trusting or trustworthy: the dominance of $\bar{C}$ makes it the obvious rational choice of both players.

Having illustrated the exercise of deterrent threats in two games, I will next formalize the concept of threat power. But I emphasize that, unlike moving or order power, in which play starts from an initial state, I assume that, when threats are made, there is

> **Prior communication (PC):** The player with threat power, in making a compellent or deterrent threat, communicates its willingness and ability to stay, if necessary, in a Pareto-inferior state.

Given PC, however, and the choice of strategies in light of PC, I assume that players can still make moves and countermoves from an initial state.

By communicating its intention to suffer, along with its opponent, at some Pareto-inferior state, the threatener who possesses threat power can induce its opponent to choose a preferred state, which the threatener also prefers. Thus, the United States can threaten to choose strategy A in game 30 (figure 5.4), and either player can threaten to choose $\bar{C}$ in PD (figure 5.2), both of which give their opponents payoffs of 1 or 2. In this situation, it is rational for the opponent to choose its strategy associated with (3,3) when the threatener also makes this choice.

All the calculations of TOM developed so far assume that players are farsighted and fully anticipate the sequence of moves and countermoves that players might make in a game. However, these calculations presuppose the single play of a game (including one which may cycle): while permitting sequential moves and countermoves by the players within a game after they have made their initial strategy choices, they do not assume that the game is repeated.

Hessel and I (Brams and Hessel, 1984) altered this assumption to permit repeated play of games.[17] Repeated play here does not necessarily mean that exactly the same game is repeated again and again – which is an assumption I criticized as unrealistic in chapter 1 – or even that it is played against the same opponent. Rather, it means that there is always later play that enables a threatener to recoup losses it may have incurred earlier in carrying out threats. Carried-out threats, I assume, enhance a threatener's credibility, enabling it to deter future challengers.

Threats, as I shall show, affect NMEs, sometimes reinforcing them and other times undermining them. Because they are intended to influence future play, they presume continuing, if not repeated, play. As Moulin

[17] The remainder of this section and the next are adapted from Brams and Hessel (1984) with permission.

(1986, p. 248) put it, "No threat is meaningful when there is no tomorrow."

To be sure, if the threat is nuclear, there may be no tomorrow once the threat is carried out. Thus, as I have argued elsewhere (Brams, 1985b, chapter 1), the credibility of nuclear threats must rest on the perception that they will be carried out, though not necessarily with certainty but with a high enough probability to make the expected payoff of $\bar{C}$ less than C (in a cardinal-utility model). When there may be no tomorrow, this perception requires that the threatener precommit itself to command and control procedures that ensure retaliation will occur sufficiently often.

By contrast, in defining threat power, I assume that the players, after a threat has been carried out, live to see another day (in the repeated game or continuing play) in which the threatener, by virtue of its action, has enhanced its credibility in the future. In the formal language to be used later, the threatener is perceived to have "threat power," which is the ability and willingness to carry out threats. This power, if correctly perceived, obviates the need actually to carry out threats.

By retaining the previous rules of sequential play in individual games, but permitting the games themselves to be repeated or play to continue, one can analyze both the stability of outcomes in repeated play and the effectiveness of threats. Thus, the ability of players to "stretch out" their calculations in anticipation of playing future games may destabilize outcomes stable in the single play of a game, based on TOM, much as myopically stable outcomes like Nash equilibria may be rendered unstable when the rules allow moves and countermoves from initial states. Likewise, states unstable in the single play of a sequential game may be rendered stable when the game is repeated, because players – anticipating its recurrence – change their rationality calculations, in ways to be made precise shortly.

Although repeated games and supergames (infinitely repeated games) have been extensively studied in economics and political science (Shubik, 1982, 1984; Ordeshook, 1986; Kreps, 1990; Friedman, 1990; Gibbons, 1992; Bierman and Fernandez, 1993), there is no comparable repeated-game analysis of games played according to TOM. Yet play under TOM rules seems ubiquitous, as I have argued, and repeated or continuing play under these rules may be just as common. The willingness of parties, for example, to accept prolonged stalemates, to refuse to negotiate at all, or to resort to the use of force – all at a considerable cost – can often only be explained by their expectation of possibly having to face the same situation over and over again. In this context, setting a precedent of implacable firmness may, though costly at the moment if challenged, more than pay for itself later on by deterring future untoward actions of opponents.

In international politics, for example, a superpower like the United States might anticipate that it will be continually engaged in the same kind of dispute with several small countries, but each small country anticipates it will have only one encounter of this kind with the superpower. If this is the case, the superpower must worry about its bargaining reputation, whereas the small country has no such concern because its stake is not continuing.[18]

Call the continuing player (superpower) the *threatener*, and the non-continuing player (small country) the *threatenee*. The threatener can make its threat credible by ignoring what it would lose in the short run if it were forced to carry out its threat.

There is, I assume, always a cost to the threatener (as well as the threatenee) when the threatener carries out its threat. If this were not the case, there would be no need to threaten an action – it would be rational simply to carry it out.

Although carried-out threats may enhance the credibility of the threatener's future threats in repeated or continuing play of a game, this is not always the case. In fact, most of the analysis in the remainder of this section is devoted to showing what kinds of threats produce what kinds of effects.[19] This analysis is also applicable to situations in which both players can threaten each other, but one has greater ability to hold out against threats – should they be carried out – and both players recognize this greater ability on the part of one.

As I noted earlier, "repeated play of a game," unlike in the standard

[18] In the Bible, the continuing player is, of course, God. That He is obsessively concerned with His reputation, which is cemented by threats that He frequently makes and carries out (Brams, 1980), is strongly related to His immortality (Brams, 1983). A number of analysts have proposed game-theoretic models that take account of bargaining reputations; these models are reviewed in Wilson (1985, 1989). An interesting game-theoretic analysis of "stability by threats" and deterrence, which mirrors some ideas developed here, is given in Moulin (1986, chapter 10); on two-person games, see Moulin (1981). Ho and Tolwinski (1982) develop a leader–follower model of a supergame, in which the leader, as threatener, establishes credibility by carrying out threats against the follower. Because this model is probabilistic, it is rational for the leader to carry out threats in some but not all instances, which is also a major conclusion of the national-security models developed in Brams and Kilgour (1988). Langlois (1991) uses a model of discounted repeated games, based on PD and Chicken, to analyze the credibility of threats in nuclear deterrence; more generally, see Powell (1989). Cioffi-Revilla (1983) offers a probabilistic model of credibility, and others (Allan, 1982; Maoz, 1983) have incorporated the related concept of "resolve" in their models. My interest, by contrast, is not in how players establish their credibility but, given that one player can make credible threats (defined precisely later), what, if any, advantages this ability confers on that player. Thus, a player's goal is not to be credible as such but to achieve its best possible outcome in games. Depending on the game, credible threats may or may not be helpful.

[19] This information seems especially important in international politics. As Baldwin (1989, p. 57) observes, "The lack of a science of threat systems is both intellectually undesirable and politically dangerous ... to world peace."

theory, may involve the threatener against either the same or different threatenees (as in the superpower–small country example). Whether both players are the same or different in repeated play, I assume that both know that the threatener is willing and able to carry out its threats to establish its future credibility.

Insofar as the threatener establishes credibility by carrying out threats, this credibility is likely to extend to repeated or continuing play of different games. Thus, a threatener who has built up a reputation for "toughness" by taking reprisals against opponents who have not met its terms can be assumed to be a qualitatively different player from one who is out to ensure its best possible outcome in any single play of a game based on TOM. Similarly, the threatenee, aware of the reputation of the threatener, will also be a different player.

Put another way, each player will have goals different from those assumed of players in the single play of a game. This requires that one specify each player's rational calculus, given the assumption of repeated play for one player (i.e., the threatener) and knowledge by the other player (the threatenee) of the threatener's concern for credibility in future play.

Assume that one of the two players, before play of a game with PC, faces repeated play of the game. Call this player T (threatener). The other player, $\bar{T}$ (threatenee), is concerned with only the single play of the game; $\bar{T}$ is aware, however, of T's continuing involvement in repeated play of the game.

As noted earlier, recurrence of a game may make it rational for T to threaten $\bar{T}$ and carry out its threat, even if it results in a worse payoff for both players when the game is played only once. To analyze the effect of repeated or continuing play of a game on the stability of the game's outcome, make the following assumption about T's threat behavior: T will stay at, or move to, a strategy disadvantageous to itself (as well as $\bar{T}$) in a single play of a sequential game iff (if and only if) this choice enables it to establish the credibility of its threats in future play of games (ultimately leading to more desirable outcomes for T). "Credibility," along with other concepts I shall discuss informally here, will be formally defined later in this section.

T's credibility, I assume, prevents $\bar{T}$ from moving to a state in a game that is advantageous to $\bar{T}$ but disadvantageous to T. Should $\bar{T}$ contemplate such a move, T's threat will be to terminate the game at an outcome disadvantageous to both players.

T's capability and resolve to carry out threats I call its *threat power*. This is the power to hurt (Schelling, 1966; Aumann and Kurz, 1977), which T may threaten to use, but will actually use, only if its threats are

	$\bar{T}$	
T	(a_{11},b_{11})	(a_{12},b_{12})
	(a_{21},b_{21})	(a_{22},b_{22})

Key: T = threatener; $\bar{T}$ = threatenee
(a_{ij},b_{mn}) = (payoff to T, payoff to $\bar{T}$)

Figure 5.5 Ordinal 2×2 payoff matrix

ignored. These threats are given force by threat power, which enables T to terminate a game at a mutually disadvantageous outcome. Although such termination is irrational for T (as well as $\bar{T}$) in any single play of the game, it becomes rational if T's involvement in a game continues and the carried-out threat deters future challenges.

To explore threats more systematically, consider the 2×2 game shown in figure 5.5, in which each player is assumed to be able to rank the four states from best to worst.[20] Assume that T desires to implement an outcome in row i and column j, (a_{ij},b_{ij}), for some $i=1$ or 2 and for some $j=1$ or 2, as the outcome of the game, but it thinks that $\bar{T}$ might move from (a_{ij},b_{ij}). Suppose, further, that to deter $\bar{T}$ from making this move, T threatens to force the termination of the game at some outcome $(a_{mn},b_{mn}) \neq (a_{ij},b_{ij})$. Call T's threat *real* (against $\bar{T}$) iff, when carried out, it worsens the payoff for $\bar{T}$. T's threat is *rational* (for itself) iff, when successful in deterring $\bar{T}$, it improves T's own payoff over what it would be if $\bar{T}$ moved from (a_{ij},b_{ij}). A threat that is both real and rational is *credible* if it satisfies certain additional conditions.

To determine the conditions under which T has credible threats, suppose, without loss of generality, that R is T. Suppose, further, that R threatens termination of the game at (a_{mn},b_{mn}) in order to prevent C from moving from (a_{ij},b_{ij}). Clearly, this implies that $b_{ij}<4$ – that is, (a_{ij},b_{ij}) is not C's best state, for if it were, C would have no incentive to depart from it, making R's threat superfluous. Now R's threat is real iff $b_{mn}<b_{ij}$; it is rational iff $a_{mn}<a_{ij}$. Combining these two inequalities, a necessary (but not sufficient) condition for R's threat to be credible is that there exists a Pareto-inferior state,

$$(a_{mn},b_{mn})<(a_{ij},b_{ij}), \tag{5.1}$$

[20] Although this matrix is not needed to follow the subsequent analysis, it may be useful to substitute the numerical subscripts in the matrix for the lettered ones in the text to illustrate the analysis.

which means that the state on the left-hand side of inequality (5.1) is worse for both players than that on the right.

If T has a credible threat, $\bar{T}$ would not move from (a_{ij},b_{ij}) only to accept Pareto-inferior state (a_{mn},b_{mn}) as the outcome. For this reason, call (a_{mn},b_{mn}) T's *breakdown state*: it forces $\bar{T}$ to comply with T's threat to avoid such an outcome, from which T will not depart. Instead, $\bar{T}$ is encouraged to stay at Pareto-superior state (a_{ij},b_{ij}), which is called T's *threat state*; T's strategy associated with this state is called T's *threat strategy*. By comparison, T's strategy associated with its breakdown state is its *breakdown strategy*.

Because T will not move from breakdown state (a_{mn},b_{mn}) – even though it is worse for both players than threat state (a_{ij},b_{ij}) – it is necessarily the outcome if $\bar{T}$ departs from (a_{ij},b_{ij}). The cost to T of suffering a breakdown state, I assume, is the price it is willing to pay in any single play of a game to ensure that its threat will not be viewed as empty (i.e., a bluff) and will be credible in future games. Henceforth, I assume that T's possession of a credible threat implies that it is *nonempty* – T will exercise it by moving to, or not moving from, its breakdown strategy.

Two qualitatively different kinds of credible threats need to be distinguished, one requiring no move on the part of T (assumed to be R in the subsequent analysis), the other requiring that T switch strategies:

1 R stays: threat and breakdown strategies coincide. When $m=i$, R's threat strategy and breakdown strategy are the same. Hence, R can threaten to stay at the Pareto-inferior state (a_{in},b_{in}) should C move there. Inequality (5.1) becomes

$$(a_{in},b_{in})<(a_{ij},b_{ij}),\ n\neq j, \tag{5.2}$$

so the existence of a state Pareto-inferior to the other state in the same row is necessary for R to have a credible threat. If, in addition, $a_{ij}\geq 3$, (5.2) is sufficient. Note also that R can always implement its threat state by choosing its threat strategy initially.

Game 30 in figure 5.4 illustrates this case. *SU* can implement its threat state, (3,3), by choosing *W* initially and threatening not to move from (2,2), *US*'s breakdown state, should *US* choose *A*. On the other hand, such a threat by either player undermines (3,3) in Chicken in figure 5.3. For example, if *SU* chooses *M* and does not budge from this strategy, it can induce (2,4) if it has threat power. Similarly, *US* can induce (4,2) with threat power, making threat power effective in Chicken because it helps the player who possesses it.

2 R would move: threat and breakdown strategies differ. When $m\neq i$, R's threat and breakdown strategies are different. In this case, R can threaten to move to its breakdown strategy should C move from (a_{ij},b_{ij}). Since C can move subsequently, it can choose as the breakdown state (a_{mn},b_{mn}), the better of its two states associated with R's breakdown strategy, which precludes $b_{mn}=1$. Necessarily, then, $b_{mn}\geq 2$. But I showed earlier that $b_{ij}<4$, and from inequality (5.1), $b_{ij}>b_{mn}$. Taken together, these inequalities imply $b_{ij}=3$ and $b_{mn}=2$. Hence b_{mk} $(k\neq n)$, C's ranking of the other state associated with Rs breakdown strategy, must be 1. In other words, C's two worst outcomes are associated with R's breakdown strategy m.

Thus, in case 2, R can threaten, should C move from its strategy associated with (a_{ij},b_{ij}), to force C to choose between its next-worst and worst states (2 and 1) by switching to its breakdown strategy m. As in case 1, R can always implement its threat state by choosing its threat strategy initially; unlike case 1, however, the Pareto-inferior state that establishes the credibility of R's threat is associated with R's other (i.e., breakdown) strategy. In sum, whereas breakdown and threat strategies are coincidental in case 1, they are not in case 2, necessitating a move on the part of R in case 2 to demonstrate that its credible threat is nonempty.[21]

Game 30 (figure 5.4) also illustrates case 2. Because *SU*'s two worst payoffs (1 and 2) are associated with *US*'s strategy of *A*, this is the breakdown strategy with which *US* can threaten *SU*. Moreover, *US* has reason to do so: it prefers its better state associated with *B*, (3,3), to *SU*'s better state associated with *A*, (2,2). Thus, (2,2) is *US*'s breakdown state, and (3,3) its threat state. As *T*, then, *US* can implement (3,3) under threat that if *SU* moves to (1,4), *US* will move to (4,1) and not move from *A*, its breakdown strategy. Although *SU* can escape its worst payoff by moving to (2,2), clearly it (as well as *US*) would prefer (3,3).

In Schelling's (1966) terms, case 1 describes a *compellent threat*: T compels $\bar{T}$ to accept the threat state by saying that it will not abandon its threat/breakdown strategy should $\bar{T}$ challenge its threat (i.e., depart from the threat state and move to the breakdown state). Case 2 describes a *deterrent threat*: T deters $\bar{T}$ from moving from the threat state by saying that it will retaliate by switching to its breakdown strategy should $\bar{T}$ challenge its threat. The two types of threats are not mutually exclusive and, indeed, may yield different threat outcomes in the same game. I shall show shortly that whenever the two threat states differ, T will prefer the one associated with its deterrent threat, assuming that both threats are credible.

[21] While it is possible to provide formal conditions under which R has a reason to threaten C, they are not very enlightening. Their significance is mostly algorithmic, and they can easily be deduced from the algorithm for determining threat states that I give at the end of this section. That rational behavior is better thought of as algorithmic rather than axiomatic is argued in Binmore (1990, pp. 153ff).

The foregoing analysis allows one to state some propositions about threat strategies and outcomes in 2×2 strict ordinal games. The first follows immediately from inequality (5.1):

Proposition 5.1. *If a game contains only Pareto-optimal states, neither player has any threat strategies. In particular, neither player can credibly threaten its opponent in a total-conflict game.*

Thus, the Pareto-inferiority of at least one state is necessary for the effective exercise of threat power in a game. Providing that the sufficient conditions given under cases 1 and 2 are met, a Pareto-inferior state becomes a breakdown state for *T*.

Proposition 5.2. *T can always implement its threat state in a game with Pareto-inferior states.*

Recall that by initially choosing its threat strategy, *T* can force $\bar{T}$ either to choose between *T*'s threat state and its breakdown state (case 1); or to risk *T*'s choosing its breakdown strategy, containing $\bar{T}$'s two worst payoffs, unless $\bar{T}$ accedes to *T*'s threat state (case 2). In either event, *T*'s threat state is Pareto-superior to its breakdown state, so $\bar{T}$ will, if rational, choose its strategy associated with *T*'s threat state at the start.

Proposition 5.3. *If T's compellent and deterrent threat states differ, T always prefers its deterrent threat state.*

To see this, note that if compellent and deterrent threat states are different, they must appear in different rows (or columns) of the payoff matrix. It follows immediately that the compellent threat state, which both players prefer to the other state in the same row (or column), must be the breakdown state of the deterrent threat; consequently, inequality (5.1) requires that it must be Pareto-inferior to the deterrent threat state.

There are exactly two games (game 48 in figure 4.6 and game 50 in figure 4.7) in which a player has different (credible) compellent and deterrent threat states. This player is *C* (*P*) in figure 4.6 and *R* (United States) in figure 4.7. The deterrent threat states of each, (3,4) in figure 4.6 and (4,3) in figure 4.7, are Pareto-superior to their compellent threat states, (2,3) in figure 4.6 and (3,2) in figure 4.7, where $\bar{B}$ is its breakdown strategy in figure 4.6 and $\bar{L}$ in figure 4.7.[22]

The following algorithm can be used to determine whether a particular 2×2 strict ordinal game has one or more threat states, (a_{ij},b_{ij}), for *R* and, if so, what they are:

1 Select *i* and *j* such that $a_{ij}=4$.
2 If $b_{ij}=4$, stop: no threats are necessary to implement (a_{ij},b_{ij}).
3 If $b_{ij}=1$, go to 8: no threats can induce (a_{ij},b_{ij}).
4 Select (a_{mn},b_{mn}) that satisfies (5.1). If none exists, go to 7.

22 In the Appendix, I indicate as the threat states in these two games only the Pareto-superior deterrent threat states.

5 If $m=i$, (a_{ij},b_{ij}) is R's compellent threat state.
6 If $m \neq i$, and $b_{mn}=2$, stop: (a_{ij},b_{ij}) is R's deterrent threat state.
7 If $a_{ij}=3$, stop: R has no threat states.
8 Select i and j such that $a_{ij}=3$. Go to 2.

When the payoffs, a and b, are interchanged, the same algorithm can be used to obtain C's threat states.

Of the 57 2×2 conflict games,

- neither player has a threat strategy in 11 (threats impotent);
- one player has a threat strategy in 26 (threats one-player potent);
- both players have a threat strategy in 20 (threats two-player potent).

The threat states that each player can induce are shown in the Appendix. Only when both players have threat strategies, and these strategies enable them to induce different states (15 of the 20 games in which threats are two-player potent), do we say their threat power is *effective*: it is better for a player to have threat power than not, because in these games one player can induce a better outcome for itself when it has threat power than when its opponent does.

In the other 5 two-player potent games, each player, by exercising threat power, induces the same outcome, making such power irrelevant, as I showed was the case in both game 30 and Prisoners' Dilemma (game 32) earlier in this section. To recall, in game 30 (figure 5.4), either a compellent threat by *SU*, or a deterrent threat by *US*, induces (3,3); in PD (figure 5.2), a deterrent threat by either player induces (3,3).[23]

5.6 The use of threat power in Poland, 1980–1

To illustrate a game in which threat power is effective, I shall use game 22 to model the conflict between the Polish Communist party and the independent trade union, Solidarity, in the 1980–1 period. In this game, if either player possesses threat power, it can induce a better state than if the other player possesses it, making threats both two-player potent and effective. Solidarity, as I shall show, was successful in asserting its influence early, but the party recouped its dominant position later.

The conflict between the leadership of Solidarity and the leadership of the Polish Communist party/government/state (assumed to be the same for present purposes) was rife with threats and counterthreats in 1980–1. Each protagonist sought to deter future challenges to its position by

[23] Earlier analyses of threat power, and comparisons with other kinds of power, can be found in Brams (1983), Brams and Hessel (1984), and Brams (1990). Sufficient conditions under which compellent and deterrent threats can be used by the threatener to induce its best outcome in two-person games larger than 2×2 are given in these sources.

seizing the upper hand, even when this proved temporarily costly. Indeed, the Polish conflict illustrates a game in which deterrent threat and compellent threats were not only effective, but both kinds of threats also came into play, albeit at different times.

Although the threat of Soviet intervention was certainly a factor in the Polish game, I will not introduce the Soviet Union as a separate player. Soviet preferences, it seems, essentially paralleled those of the Polish Communist party. Hence, the Soviets need not be modeled as a separate player but rather can be treated as a force on the side of the party that affected the balance of power in the game and, consequently, the eventual outcome.

Each of the two sets of leaders may be treated as if it were a single decision maker. Of course, internal divisions within Solidarity and the party led to certain intra-organizational games; however, these subgames generally concerned not strategic choices on broad policy issues but rather tactical choices on narrower operational questions. Focusing on the main game has the advantage of highlighting the most significant political–military choices each side considered, the relationship of these choices to outcomes in the game, and the dependence of these outcomes on threats and threat power.

The two players faced the following choices:

1 **Party.** Reject or accept the limited autonomy of plural social forces set loose by Solidarity. Rejection would, if successful, restore the monolithic structure underlying social organizations and interests; acceptance would allow political institutions other than the party to participate in some meaningful way in the formulation and execution of public policy.

2 **Solidarity.** Reject or accept the monolithic structure of the state. Rejection would put pressure on the government to limit severely the extent of the state's authority in political matters; acceptance would significantly proscribe the activities of independent institutions, and Solidarity in particular, to narrower nonpolitical matters, with only minor oversight over certain state activities.

I designate these strategies of both sides as "rejection" (R) and "acceptance" (A). They might also be designated "confrontation" and "compromise," but I prefer the former more neutral labels because the disagreements were generally over specific proposals rather than general postures that the two sides struck.

The two strategies available to each side give rise to four possible states:

1 ***A–A*.** Compromise that allows plural institutions but restricts their activities to nonpolitical matters, with negotiations commencing on the sharing of political power.

Solidarity \ Party	Rejection (*R*)	Acceptance (*A*)
Acceptance (*A*)	Status quo ante (2,4)* [2,4]	Compromise, with restrictions on political activities of Solidarity (3,3)# [2,4]
Rejection (*R*)	Violent conflict, with possibility of outside intervention (1,2) [2,4]	Authorization of independent political activity, leading to gradual reduction of the party's role (4,1) [2,4]

↑ Dominant strategy (Party: Rejection (*R*))

Key: (x,y) = (payoff to Solidarity, payoff to Party)
$[x,y]$ = [payoff to Solidarity, payoff to Party] in anticipation game (AG)
4 = best; 3 = next best; 2 = next worst; 1 = worst
Nash equilibrium in original game underscored
NME circled
* = Party's compellent threat state
\# = Solidarity's deterrent threat state

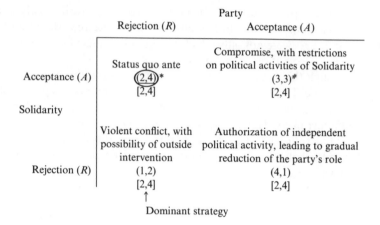

Figure 5.6 Polish crisis, 1980–1 (game 22)

2 ***R–R.*** Possibly violent conflict involving the entire society, opening the door to outside (mainly Soviet) intervention.
3 ***A* (Solidarity)–*R* (Party).** Status quo ante, with tight restrictions on all activities of Solidarity and its recognition of the supremacy of party/state interests.
4 ***R* (Solidarity)–*A* (Party).** Authorization of independent political activity, and a corresponding gradual reduction of the party/state role in implementing public policy decisions made collectively.

I next justify the rankings I assign to the different states, which are shown in the payoff matrix of figure 5.6.

The party leadership repeatedly emphasized the unacceptability of any solution that would constrain its political power, which implies that its two worst outcomes are those associated with Solidarity's choice of *R*. Commenting on the Polish events, Bialer (1981, p. 530) wrote, "Some [Party] leaders are more conservative and some more reformist; none, to our knowledge, questions the need to preserve the Party's monopoly." The Eleventh Plenary Meeting of the Central Committee of the Polish United Workers Party (PUWP) was explicit on this point:

> The Central Committee of PUWP unequivocally rejects ... concepts of abandoning the leading role of the Party, of reducing this role to the ideological sphere and dispossessing the Party of the instruments of political power. This is the main danger. (*Nowy Dziennik*, June 17, 1981, p. 2)

In fact, the available evidence indicates that the party preferred an all-out confrontation (*R*–*R*) to relinquishing its supremacy [*R* (Solidarity)–*A* (Party)]. Speaking at the Ninth Congress of PUWP, Deputy Prime Minister Mieczyslaw Rakowski announced: "To the enemies of socialism we can offer nothing but a fight, and not merely verbal at that" (*Nowy Dziennik*, July 21, 1981, p. 2). A later declaration of the Politburo reiterated that challenge: "We shall defend socialism as one defends Poland's independence. In this defense the state shall use all the means it deems necessary" (*Nowy Dziennik*, September 22, 1981, p. 2). Finally, between its two best states associated with Solidarity's choice of *A*, the party clearly preferred the status quo [*A* (Solidarity)–*R* (Party)] to compromise (*A*–*A*).

As for Solidarity, there is considerable evidence that it preferred the party's capitulation [*R* (Solidarity)–*A* (Party)] most, and violent conflict (*R*–*R*) least. In between, it preferred a compromise solution (*A*–*A*) to its own capitulation [*A* (Solidarity)–*R* (Party)]. Solidarity statements echoed this sentiment. Its chairman, Lech Walesa, said, "We don't want to change the socialist ownership of the means of production, but we want to be real masters of the factories. We were promised that many times before" (*Time*, September 8, 1980, p. 34). Jacek Kuron, one of Solidarity's advisers, further clarified where the line on party activities should be drawn: "The Party's leading role means the monopoly of power over the police forces, the army and foreign policy. All other matters must be open to negotiations with society" (*Time*, December 29, 1980, p. 29). In short, Solidarity preferred not to try to rob the party of its most significant functions, hoping to gain the party's acquiescence and thereby at least Solidarity's next-best state (*A*–*A*).

The reason for Solidarity's preference is evident. Solidarity was aware of the unacceptability of its best state [*R* (Solidarity)–*A* (Party)] to its adversary:

> From the start, the Polish workers understood [that] to think of overthrowing the Party in Poland was madness, for it would inevitably lead to Soviet invasion and the destruction of all liberties gained in the past ten or even twenty-five years. (Ascherson, 1982, pp. 18–19)

Addressing Solidarity members, Walesa said: "Our country needs internal peace. I call on you to be prudent and reasonable" (*Time*, December 29, 1980, p. 20). On another occasion he said, "[There are] fighters who want to fight at every opportunity; but we must understand that both the society and the union have had enough of confrontation ... we ought not to go to the brink" (*Solidarnosc*, April 10, 1981, p. 1).

Kuron (1981) concurred: "The goals of the government and of the democratic movement are completely opposite. But the struggle between the two tendencies, the totalitarian and the democratic one, are to be fought exclusively by peaceful means." Thus, Solidarity preferred *R* only if the party chose *A*; if the party chose *R*, it preferred *A*.

The payoff matrix in figure 5.6 is game 22, which is one of the 12 specific games subsumed by the Magnanimity Game (MG) (figure 3.5). Because it is a noncyclic game (it was discussed in section 4.2), moving power in it is not defined. Neither is order power, because it contains no indeterminate states.

As shown in its anticipation game (AG), game 22 has only one NME, (2,4). This state is the unique Nash equilibrium in the original game, associated with the party's dominant strategy of *R*. Anticipating this choice by the party, Solidarity would choose *A* because it prefers (2,4) to (3,3), making (2,4) the apparent solution in this game.

Because this solution is the best for the party, but only next worst for Solidarity, the game seems to be inherently unfair to Solidarity. Solidarity, however, can undermine this NME if it possesses threat power. In fact, it can induce (3,3) by a deterrent threat: choosing *A*, it can threaten *R*, and the party's two worst states (1 and 2), unless the party accepts Solidarity's threat state, (3,3).

Solidarity's breakdown state, (1,2), is also the breakdown state for the party if it is the player with threat power. The party's threat is compellent and is implemented by its choosing, and staying at, *R*, forcing Solidarity to choose *A* and the party's threat state, (2,4).

Thus, with confrontation (*R*–*R*) being the common breakdown state, the game turns on which (if either player) holds the balance of power. If Solidarity is the more powerful of the two players, or is at least perceived as such, it can implement (3,3). Otherwise, (2,4), as both the Nash equilibrium and NME, would presumably obtain – and be reinforced should the party possess threat power.

Note that Solidarity can implement its threat state, (3,3), by choosing its "soft" strategy of *A*, relegating its "hard" strategy of *R* to a (deterrent) threat. This, of course, is the proverbial "speak softly and carry a big stick" policy, with the big stick being used only if necessary.

In contrast to Solidarity, the party always does at least as well as Solidarity [(3,3), if it does not have threat power], and sometimes better [(2,4) if it does, or neither player does], at least in terms of the comparative rankings of the two players. Thus, a power asymmetry unfavorable to itself is not as serious for the party as for Solidarity, again based on a comparative ranking of the two threat states. Moreover, because the party's threat is compellent, it can implement its best outcome simply by choosing and then maintaining its hard strategy of *R*, whereas Solidarity must first take a soft position and then threaten escalation to its hard position, putting the onus for breakdown and subsequent disruption on itself.

This analysis of threat power offers meaningful insights into the actual unfolding of events in Poland in 1980–1. Clearly, the party was stunned by the quick pace of developments and the widespread support for Solidarity after the August 1980 Lenin shipyard strike in Gdansk, which is the event that I take as the starting point of the game.

After this spark that set off the crisis, the party did in fact consider Solidarity to be more powerful during the last part of 1980 into the beginning of 1981. Reluctantly, it followed its acceptance strategy, *A*; Solidarity, for its part, repeatedly emphasized the nonpolitical character of its demands (*A*) while threatening *R*, for the union's very existence was "based on adversary relations, not on a partnership" (Szafar, 1981, p. 79). As the economic situation deteriorated, however, the instability of the (3,3) compromise state became evident, setting the stage for a test of strength between Solidarity and the party.

In March 1981, for the first time, the government used force against Solidarity. Although the force was limited in scope, its use can be interpreted as an attempt to switch back to the party rejection strategy of *R*. Yet Solidarity chose to avoid confrontation, and the game remained at (3,3).

Although "the game was to leave the authorities with a semblance of power but to take its substance away" (Watt, 1982, p. 11), the events of March 1981 began to split Solidarity, strengthening proponents of the rejection strategy. But the moderate leadership of Solidarity, and Walesa in particular, kept pointing to society's unwillingness to support the rejection strategy.[24] In doing so, the leadership cast doubt upon the viability of Solidarity's breakdown strategy, thereby undermining the union's power.

24 Walesa, for example, was opposed to a nationwide strike because he considered the Polish situation too precarious, both internally and externally. Defending his position, he said, "We must move step by step, without endangering anybody . . . I am not saying that there will be no confrontation over some important issue . . . but now the society is tired and wants no confrontation" (*Solidarnosc*, April 10, 1981, p. 1). More belligerently, Erich Honecker, the East German Communist party leader, urged Warsaw Pact members to invade Poland at an emergency meeting in Moscow on December 5, 1980, but this option was rejected by the Soviet Union and other members for a variety of reasons (Kinzer, 1993).

In December 1981 the party, apparently believing that the balance of power had shifted, switched decisively to its rejection strategy *R*, moving the game to its threat state, (2,4), by imposing martial law and jailing many Solidarity leaders in a massive crackdown. The relative stability of this outcome until 1989 seemed to validate the party's assessment that the balance of power in Poland favored it – or at least demonstrated that Solidarity's power was not greater than its own.

Although Solidarity showed no appetite to switch to R, by 1989 the catastrophic economic situation led the party to take a much more accommodationist position. (I have not tried to model how this situation evolved or the new preferences of the players, but the Soviet Union was no longer a significant factor.) In eight-week roundtable talks in March–April 1989 between the party and Solidarity, an agreement was reached to hold elections in June, in which Solidarity was allowed to compete for 35 percent of the seats in the lower house, and all the seats in the upper house, of parliament. In an embarrassing turn of events for the party, Solidarity not only won virtually all the contested seats, but many party members failed to win majority support even in uncontested elections. The party was further discredited when Walesa was elected president in December 1990 and completely free elections were held for both houses of parliament in October 1991.

The negotiated compromise and election results can perhaps be viewed as a return to the (3,3) state. It was not so much provoked by threats of violent conflict but rather a recognition by both sides that the (4,2) state after the December 1981 crackdown had, because of severe economic problems and revolutionary changes in other Communist countries in Eastern Europe in 1989–90, degenerated to something probably approaching (1,2). Faced with a disastrous economy and political situation, both sides saw benefit in trying to reach a *modus vivendi.*

5.7 Varieties of power

Throughout this book I have argued the need for looking at conflicts as dynamic events that unfold over time, which may uncover either stability or instability in games that the standard theory hides or places elsewhere. The variable I have added to the equation in this and the previous chapter is power. Although a central, if not the central, concept in the fields of political science and international relations, power has proved to be an elusive concept indeed, even at a purely theoretical level.

But definitions that tap different aspects of power – moving, order, and threat in this analysis – come to the fore naturally in games in which players can make rational moves and countermoves. For example, moving power focuses on the effects of one player's being able to move indefinitely, whereas order power assumes that one player is able to determine the order of play from an initial (indeterminate) state. Threat power posits prior communication (PC), enabling one player to threaten the other and carry out its threat if necessary, because the game is repeated or continuing and this loss will be recouped later.

In conflicts at all levels, an asymmetry in capabilities is probably at least as common as players having symmetric or equal capabilities. Such an asymmetry, surprisingly, may contribute to stability. The compromise in the Cuban missile crisis – if that is what it was – has remained more or less intact for more than a generation, which seems testimony to the durability of a settlement worked out between unequal players.

Although the United States had subsequent conflicts with the Soviet Union over Cuba as well as with Cuba itself, none involved a confrontation with the Soviet Union on nearly the scale that occurred in 1962. As a result of this crisis and the apprehension and fear it aroused, a hot line was established in 1963 linking the heads of state in Washington, D.C., and Moscow, which has on occasion been used to try to prevent displays of brinkmanship from carrying the parties again so close to the threshold of nuclear war.

With the breakup of the Soviet Union, this link may no longer seem necessary. But such lines of communication in the future, perhaps between different antagonists, may prove useful. Indeed, because threats need to be communicated, and in games like PD or game 30 they facilitate compromise, they may well be conducive to peace. On the other hand, threats in Chicken can undermine cooperation, so, from a policy perspective, threats need to be evaluated in the context of the game being played.

The Polish game, at least in the 1980–1 period, illustrates the effectiveness of threat power: it made a difference, depending on whether Solidarity or the party possessed it, on what outcome was implemented. Solidarity's threat was in fact credible in the beginning, and the union temporarily gained some concessions from the party. But these were not to last, because the party, with Soviet support, could also use threat power to redress the imbalance and reinstitute the Nash equilibrium (and NME) in game 22, which it eventually did.

Threats will be decisive if they are backed up by the wherewithal to carry them out. But to avoid a Pareto-inferior outcome, the threatenee must recognize that the threatener will act – even to its own detriment – if the threat is ignored. Given both the power and the resolve to implement threats (should this prove necessary), threats can serve either to deter or to compel desired behavior – as the Polish conflict in 1980–1 illustrated – which may or may not represent a compromise outcome.

The problem with threats, of course, is that the threatener as well as the threatenee will in general be hurt in the breakdown state if the threatener is forced to carry them out, thereby making them appear quite incredible sometimes. That is why it is necessary to postulate repeated or continuing play to model threats, because carrying out a threat in the single play of a game is, by definition, irrational.

The execution of threats is probably less likely when threat power is irrelevant. This was true of threat as well as moving power in the figure 5.4 version of the Cuban missile crisis (game 30). On the other hand, order and threat power are effective in the figure 5.3 version (Chicken), but I suggested in section 5.4 that the United States had good reason for backing away from exercising its apparently greater power in this crisis. Games in which power is irrelevant, or players may not wish to use it if it is effective, probably better lend themselves to amicable solutions than games in which power is effective and players relish using it, perhaps to cement their reputations for toughness.

When one player can, by using its power, implement for itself an outcome superior to that which an adversary can achieve using its power, then there will be good reason for the players to vie for influence, as in the Polish crisis. As another case in point, both moving power and order power are effective in the game 56 model of the Samson and Delilah story, though they predict different outcomes when Delilah is the player with such power.

If there is a problem with the analysis of power within the framework of TOM, it lies in its lack of economy: because of the varieties of power, there are different predictions to sort out and different dimensions of its use to understand. But I believe this problem is as much tied to the inherent complexity of the phenomena being studied (i.e., varieties of power) as to any large gaps in the theory.

Doubtless, simplifications can and will be made, perhaps making the theory more crisp and elegant, but they should not be at the expense of concealing aspects of power that are significant and observable in the world today. Power is multifaceted, and to model its exercise in different strategic situations may require a careful choice from among different concepts. Of the three varieties I have analyzed, one seems likely to be applicable in most situations in which there is an imbalance in capabilities.

6 Information in games: misperception, deception, and omniscience

6.1 Introduction

The classical definition of a game in extensive form includes the specification of the information players have about the sequential moves of other players. Although the role of information in games has always been central in game theory, it takes on new dimensions under the rules of play of TOM.

For example, because the standard theory pays little attention to possible differences in the power of players,[1] it also has little to say when there is incomplete information about these differences. Yet a player's information about the power of its opponent is often as significant as its information about its opponent's preferences. In this chapter, I shall explore some consequences of both kinds of information, or the lack thereof, on the play and outcomes of games.

With the exception of North Vietnam's misperception of U.S. preferences at the end of the Vietnam war (section 4.5), I have assumed until now that players plot their moves and countermoves knowing each other's preferences for the different states. However, many games are not ones of complete information. As I shall show, President Jimmy Carter's attempt to gain release of the U.S. embassy hostages who were taken and held by Iran in 1979–81 illustrates not only an embarrassing lack of information on his part but also, as in the case of North Vietnam in 1972, misinformation about his antagonist's preferences.

There may also be incomplete information about the relative power of two players. Assume that a threatener (T) has the ability and will to carry out a threat and that its threat is credible in the game being played (section 5.5). Furthermore, assume there is prior communication (PC), so the threatenee ($\bar{T}$) knows the intentions of T when $\bar{T}$ is made the object of a compellent or deterrent threat. Still, $\bar{T}$ may not recognize that T

[1] Powell (1989) defines the "resolve" of players to use threats in his models of nuclear deterrence, which may be considered as one indicator of power.

possesses threat power and, especially if this power is effective, $\bar{T}$ may attempt to defy T.

In such a case, incomplete information about the possession of threat power may lead to a fight, resulting in a breakdown state worse for both players than the threat state. Likewise, if there is not common knowledge about which player, if either, possesses moving power, cycling may ensue until one player is eventually forced to stop. With order power, too, incomplete information about which, if either, player possesses it at an indeterminate state may lead each to try either to outlast the other at that state or to move away first, depending on whether moving first or moving second is more beneficial to a player.

In section 6.2, I shall illustrate the effects of incomplete information about threat power by invoking game 22 again, which was used in section 5.6 as a model of the Polish crisis in 1980–1. But this time I use game 22 to model the confrontation between the Union and the Confederacy just prior to the outbreak of the American Civil War in 1861.

In this crisis, some leaders of the Confederacy appeared to understand quite well its inferior strategic position. Nevertheless, they attempted to coerce Union leaders into believing, through both threats and actions, that the Confederacy possessed threat power. Despite early military successes, however, the Confederacy was eventually brought to its knees by a long and bloody war.

If the South had had better information about the resources and resolve of the North, conceivably the war could have been avoided or at least settled more quickly. By 1861, however, the issue of slavery could no longer be sidestepped or postponed, as it had been by the Missouri Compromise in 1821 and the Compromise of 1850.

Which side could enforce its will on the other, once threats had failed to produce another compromise, was by no means apparent. If the strategic situation were less murky about who had threat power, as it seems to have been in the Polish crisis, then a test of strength might not have been needed.

In section 6.3 I show how the effects of incomplete information can be explored more systematically in the generic Magnanimity Game (MG) (section 3.4), in which players are assumed to have no information about the preferences of their opponents. I derive two propositions about what information each player, not knowing the type of opponent it faces, would have to acquire to make a rational strategy choice in MG.

In section 6.4 I analyze the confrontation between Jimmy Carter and Ayatollah Khomeini over the release of the American hostages in Iran that I alluded to earlier. Carter erroneously assumed that he knew Khomeini's preferences. Khomeini, it seems, did not willfully deceive

Carter; rather, the U.S. president misestimated Khomeini's position and, as a consequence, bungled the crisis.

On the other hand, I show in section 6.5 that Premier Khrushchev may have consciously deceived President Kennedy about his preferences in the Cuban missile crisis (section 5.4). However, his deception – or perhaps a sincere change in preferences – facilitated a resolution to the crisis. Indeed, there is a set of games in which, paradoxically, it is in the interest of one player to be deceived by the other.

Shakespeare's *Hamlet* illustrates a situation in which the acquisition and selective dissemination of information is crucial to the outcome. I analyze the conflict between Hamlet and Claudius in section 6.6, where I argue that Hamlet's dithering is perhaps less a tragic flaw in his character than a rational response in a game in which sketchy information had first to be filled in and verified before rational action could be taken.

There is a set of games, discussed in section 6.7, in which being omniscient is a curse rather than a blessing. This odd result can also be interpreted as illustrating the power of commitment, akin to that exercised in making a compellent threat. Here, however, I stress how a player's awareness that its omniscient opponent can anticipate its choices may, anomalously, hurt the omniscient player.

I conclude with some thoughts on the double-edged effects of information. Although it is usually helpful to know more rather than less about an opponent and its capabilities, in certain games ignorance – or at least a lack of omniscience – may be bliss.

6.2 Was the Civil War a result of incomplete information?

To illustrate the effects of incomplete information on threat power, consider the game 22 model of the Polish crisis in 1980–1, discussed in section 5.6. I indicated that this crisis did not erupt in civil war for two reasons: (1) each side recognized the dire consequences of such a war; and (2) the two sides held sway at different times in the crisis, with each well aware when it could resist (because it would not be challenged) and when it must concede (because it would). Because both sides did not pursue their threat strategies against each other at the same time, the breakdown outcome in this game was avoided.

Specifically, the party deferred to Solidarity early in the crisis, whereas Solidarity deferred to the party later, when its options were severely limited. Thus, for example, after martial law was declared, Solidarity had the choice of violent or passive resistance. Recognizing the futility of the former course, it chose instead to wage a nonviolent campaign of resistance.

By contrast, there was no such mutual recognition of the balance of forces in the United States in 1861. The overriding issue, of course, was the institution of slavery, and the two sides were the Union and the Confederacy.[2]

In my brief recounting of the game played just before the outbreak of the Civil War, I shall argue that the preferences of the two players were exactly the same as those in the Polish crisis. However, the outcome of game 22 this time was the breakdown state, in part, it seems, because the Confederacy refused to recognize the threat power of the Union.

But there is also evidence that at least some Confederate leaders understood that the South was in a decidedly inferior position. Nevertheless, they chose to escalate the crisis, but not with the expectation of committing political suicide. Although they had little hope of winning a civil war – if it came to that – they thought that by demonstrating their unswerving commitment to secede, they could convince the North to back down, once it understood the depth of their commitment and the costs the Union would incur in trying to put down an insurrection by the Confederacy.

Before showing the crucial role that incomplete information about the balance of forces played, let me first justify the game. Each side could choose between a compromise and a noncompromise strategy:

1 **Union.** Compromise (C) would mean permitting the Confederacy to keep their slaves and then negotiating their allowance in certain other territories, whereas $\bar{C}$ would mean demanding the immediate abolition of slavery.

2 **Confederacy.** Compromise (C) would mean accepting restrictions on slavery outside the South, whereas $\bar{C}$ would mean demanding the right to hold slaves in the United States or, more likely, seceding from the Union and maintaining slavery in the South, probably as an independent country.

The preferences of the players for the four possible states, moving clockwise from the upper left-hand state in figure 6.1, are as follows:

I **Compromise: (3,3).** A negotiated settlement, whereby slavery would be allowed in the South but restrictions would apply elsewhere, which is next-best for both players.

II **Union submits: (1,4).** Slavery is permitted in the United States – or, alternatively, the Confederacy secedes without a civil war – which is worst for the Union and best for the Confederacy.

III **Civil war: (2,1).** Negotiations fail because neither side backs down, and a major war ensues. This is next-worst for the Union, on the

[2] This section is adapted from a student paper of a former NYU undergraduate, Brian J. Lahey.

		Confederacy		
		Compromise (C)	Don't compromise ($\bar{C}$)	
Union	Compromise (C)	I Compromise (3,3)* [4,2]	II Union submits (1,4) [4,2]	
	Don't compromise ($\bar{C}$)	IV Confederacy submits (4,2)# [4,2]	III Civil war (2,1) [4,2]	←Dominant strategy

Key: (x,y) = (payoff to Union, payoff to Confederacy)
$[x,y]$ = [payoff to Union, payoff to Confederacy] in anticipation game (AG)
4 = best; 3 = next best; 2 = next worst; 1 = worst
Nash equilibrium in original game underscored
NMEs circled
* = Confederacy's deterrent threat state
# = Union's compellent threat state

Figure 6.1 Union–Confederacy conflict (game 22)

expectation that it will prevail in the end – albeit at a high price – and worst for the Confederacy on the expectation that it will be defeated and slavery will be abolished.

IV **Confederacy submits: (4,2).** Slavery is abolished without civil war, which is best for the Union but next-worst for the Confederacy.

As can be seen from figure 6.1, the Confederacy always prefers that the Union choose C (4 and 3) rather than $\bar{C}$ (2 and 1). Given the Union's choice of C (first row), the Confederacy would prefer $\bar{C}$ (4) to C (3) – that is, having the Union submit rather than compromise.

But what if the Union chooses $\bar{C}$, which is, after all, its dominant strategy? I presume in the ranking that the Confederacy prefers C (2) to $\bar{C}$ (1) – submission rather than civil war – which is supported by the observation of a contemporary observer:

> The truth was, the southern disunionists did not wish war, and they did not believe it would happen. The state of their finances would not sanction it, to say nothing of the dubious result of a collision with the colossal power of the north, backed by her navy. (Headley, 1863, pp. 52–3)

Just as the Confederacy most desired compromise on the part of the Union, the Union had a similar desire that the Confederacy choose C (4 and 3), with an obvious preference for submission by the Confederacy (4) over compromise (3). But unlike the Confederacy, the Union did not prefer C over $\bar{C}$ if the Confederacy was adamant (chose $\bar{C}$): "Let there be no compromise on the question of *extending* slavery," Abraham Lincoln wrote to Lyman Trumbull on December 10, 1860 (cited in Staudenraus, 1963, p. 52; italics in original). As another indicator of Union steadfastness on the slavery issue, Congressman Oris S. Ferry of Connecticut advised Gideon Welles on December 11, 1860, that ten years of civil war would be preferable to a division of the Union (Stampp, 1950, p. 28).

Given the unique NME of (4,2) in game 22, which is also the Nash equilibrium supported by the Union's dominant strategy of $\bar{C}$, why was this state not chosen by the players? One plausible reason is that some Confederacy leaders thought that their deterrent threat of choosing $\bar{C}$, which includes the two worst states (1 and 2) of the Union, could induce the compromise of (3,3). To be sure, the Union's compellent threat of choosing $\bar{C}$ induces (4,2), which is best for the Union, so the question of which side prevails in a "threat contest" obviously depends on which side possesses threat power.

Although a lack of information about who possessed such power is surely one explanation of why the threat contest escalated to a civil war, an alternative explanation is also persuasive: the Confederacy, recognizing its weaker position (the white population of the South was less than one-third that of the North), sought to prove that it was at least the equal of the North in other ways. By launching an attack on Fort Sumter in April 1861, and scoring early victories that it hoped would demoralize the North, a prolonged and costly war could be avoided:

> The surest way to prevent this [war] would be to make the contest appear [as] equal as possible by getting the entire south to act in unison. Then the North would shrink from the appalling evils of a civil war, and grant them their independence. (Headley, 1863, p. 53)

Thus, it seems, the bloodiest war in American history was provoked, at least in part, because the Confederacy thought it could threaten the Union into acquiescence, if not submission, with an early and effective show of force. Although the show of force was impressive, the North, under Lincoln, was not intimidated and fought doggedly on for four years, which eventually led to Sherman's devastating march through the South and the Confederacy's surrender at Appommatox in April 1865.

Perhaps many in the South would have preferred to sue for peace after the tide of battle turned in favor of the North at the Battle of Gettysburg in July 1863. But by then it was too late for the South to withdraw. When the North was finally able to assert its threat power, Lincoln was determined to make it palpably clear that the North could better withstand the breakdown state of (2,1).

The Civil War was a clash of interests over a fundamental issue.

Moreover, the differences of the two sides over slavery in 1861 were probably irreconcilable by any kind of compromise. The question the present analysis raises is not whether the opportunity to fashion a compromise was lost but rather whether better information about the strength and determination of Lincoln to force the question – by military means if necessary – could have induced a less hostile response from the South. Realistically, probably not, but the analysis highlights how the outcome of the conflict seems to have turned on which side possessed threat power, or thought it did.

I raised this kind of question earlier in analyzing moving power in section 4.3. In that section I showed that in three of the four class 1 MG games (33, 34, and 35), moving power is effective. In particular, if *D* possesses such power, but *V* does not know this, a contest may ensue to determine which player can hold out longer. I suggested that Israel's failure to appreciate Egypt's rebound capacity after each of their five wars through 1973 helps to explain their chronic nature (see also section 4.6). Only after the 1973 Yom Kippur war did Israel come to appreciate Egypt's moving power, paving the way for the 1979 peace treaty.

Although "learning the hard way" may not have been efficient in either the Egyptian–Israeli conflict or in the Civil War, it may in both cases have had the effect of creating the conditions for a more long-term solution than – in the absence of a true test of strength – trying once again to patch over differences. Without such a test, information remains incomplete about which side in fact has the greater power, even when the two sides know each other's preferences and, therefore, can make the TOM calculations I have described.

As long as the balance of power remains in doubt, it is easy for one side to overestimate its capabilities and think that it can coerce the other into submission. Like the wars between Egypt and Israel between 1948 and 1973, the decades of lower-level yet persistent conflict between the North and South on the slavery issue, beginning in 1819 over the admission of Missouri to the Union, illustrate this point. Fortunately for both sides in the aftermaths of these conflicts, the settlements that were achieved in 1865 and 1979 have proven quite sturdy.

6.3 Incomplete information in the Magnanimity Game (MG)

Not only may information about the relative power of the players be incomplete, but so may be information about their preferences. Although a player may have no information about an opponent, more likely it will have some information, such as a partial ordering of its preferences. Consider, for example, the Magnanimity Game (MG), which was analyzed in section 3.4. In this game, depicted in figure 3.4, *V*'s preferences were restricted to two types and *D*'s to six types, yielding 12 specific games (figure 3.5).

To analyze the effects of incomplete information on the play of MG, consider *V*'s two types of preferences: given *D* chooses $\bar{C}$, either *V* prefers $\bar{M}$ (Rejected Status Quo) over *M* (Rejected Magnanimity), or vice versa. Call the first type *hard* (*V* prefers to be tough when *D* is), and the second type *soft* (*V* prefers to be lenient when *D* is tough).

Assume that *V* knows whether it is hard or soft but does not know *D*'s six types of preferences (I will not give them names). If *V* is totally ignorant of *D*'s type, it is unable to choose a *rational strategy* – that is, one that always contains an NME, given that play commences at Status Quo. However, observe in figure 3.5 that if *V* is soft, which it is in games 34, 29, 50, 18, 22b, and 57, its rational strategy is *M* in all games except game 34, which defines one type of *D* opponent. If *V* is hard, which it is in games 22a, 33, 35, 28, 32, and 11, its rational strategy is $\bar{M}$ in all games except 28 and 32, which define two types of *D* opponent. These results can be summarized as follows:

> **Proposition 6.1.** *If V is soft, it need rule out only one type of D (out of six) to render its strategy of M rational. If V is hard, it must rule out two types to render its strategy of* $\bar{M}$ *rational.*

As one would expect, a soft *V* will generally choose *M* and a hard *V* will generally choose $\bar{M}$.

Now reverse matters and assume that *D* knows which of the six types it is. However, assume that *D* does not know whether *V* is hard or soft. This means that *D* can narrow down the games to pairs of which there are six, depending on *V*'s type (hard, soft):

(22a, 57); (35, 50); (33, 34); (28, 29); (32, 22b); (11, 18).

D's choice of *C* is rational in (38, 50), (33, 34), and (28, 29), whereas $\bar{C}$ is rational in (11, 18); neither strategy is rational in both games of pairs (22a, 57) and (28, 29). These results can be summarized as follows:

> **Proposition 6.2.** *Assume D knows its own preferences, of which there are six types. Four of D's six types can choose a rational strategy (either C or* $\bar{C}$*) without knowing whether V is hard or soft. For D's two remaining types, this knowledge is necessary for D to choose a rational strategy.*

The four games that represent the latter two types are Prisoners' Dilemma (game 32) and Chicken (game 57) – and their two counterparts, in which *V* has Prisoners' Dilemma preferences and *D* Chicken preferences (game 22a) and vice versa (game 22b). Thus, *D* can determine its rational strategy if its preferences are not those of either Prisoners' Dilemma or

Chicken. If they are, it must ascertain whether V is hard or soft (i.e., has Prisoners' Dilemma or Chicken-type preferences), which fixes one of the four specific games.

I shall not attempt to apply Propositions 6.1 and 6.2 to particular cases, though both might help to explain some of the different outcomes of MG that occurred in the historical examples discussed in section 3.5. What they may be even more applicable to is policy analysis, given a country is D or V after a war and must choose between two basic policy stances. Thus, if it is D, and its preferences make it one of the four types in which either C or $\bar{C}$ is always rational – whether V is hard or soft – then it obviously has a clear-cut choice. But if D is one of the other two types, it would behoove it to try to gather intelligence on V's type before choosing C or $\bar{C}$. Similarly, V's rational choice is also dependent on what type D is, but it need not identify precisely which of the six types but only ascertain that it is not at most one of two types.

I shall next give an example of a game of incomplete information in which the relative power of the players was not the main issue. Instead, it was incomplete information about preferences, which led one player to misperceive the goals of its opponent.

6.4 Misperception in the Iran hostage crisis

In the Iranian seizure of American embassy hostages in November 1979, the military capabilities of the two sides were almost irrelevant.[3] Although an attempt was made in April 1980 to rescue the hostages in an aborted U.S. military operation that cost eight American lives, the conflict was never really a military one. It can best be represented as a game in which President Jimmy Carter misperceived the preferences of Ayatollah Ruhollah Khomeini and attempted, in desperation, to find a solution in the wrong game.

Why did Khomeini sanction the takeover of the American embassy by militant students? Doing so had two advantages. First, by creating a confrontation with the United States, Khomeini was able progressively to sever the many links that remained with this "Great Satan" from the days of the Shah. Second, the takeover mobilized support for extremist revolutionary objectives just at the moment when moderate secular elements in Iran were challenging the principles of the theocratic state that Khomeini had installed.

President Carter most wanted to obtain the immediate release of the hostages. His secondary goal was to hold discussions with Iranian relig-

[3] This case is adapted from a student paper of a former NYU graduate student, Walter Mattli, and from Mattli (1992).

ious authorities on resolving the differences that had severely strained U.S.–Iranian relations. Of course, if the hostages were killed, the United States would defend its honor, probably by a military strike against Iran.

Carter considered two strategies:

1 **Negotiate (*N*).** With diplomatic relations broken after the seizure, negotiations could be pursued through the U.N. Security Council, the World Court, or informal diplomatic channels; the negotiations might include the use of economic sanctions.
2 **Intervene militarily (*I*).** Military action could include a rescue mission to extract the hostages or punitive strikes against selected targets (e.g., refineries, rail facilities, or power stations).

Khomeini also had two strategies:

1 **Negotiate (*N*).** Negotiations would involve demanding a return of the Shah's assets and an end to U.S. interference in Iran's affairs.
2 **Obstruct (*O*).** Obstructing a resolution of the crisis could be combined with feigning to negotiate.

Carter's view of the game is shown in the top matrix of figure 6.2, which is game 50.[4] He most preferred that Khomeini choose *N* (4 and 3) rather than *O* (2 and 1), but in any case he preferred *N* to *O*, given the difficulties of military intervention.

These difficulties were compounded in December 1979 by the Soviet invasion of Afghanistan, which eliminated the Soviet Union as a possible ally in seeking concerted action for release of the hostages through the United Nations. With Soviet troops next door in Afghanistan, the strategic environment was anything but favorable for military intervention.

As for Khomeini, Carter thought that he faced serious problems within Iran because of a critical lack of qualified people, demonstrations by the unemployed, internal war with the Kurds, Iraqi incursions across Iran's western border, and a continuing power struggle at the top (though his own authority was unchallenged). Consequently, Carter believed that negotiations would give Khomeini a dignified way out of the impasse (Carter, 1982, pp. 459–89).

One implication of this view is that while Carter thought that Khomeini most preferred a U.S. surrender at *NO* (4), he believed Khomeini would next most prefer the compromise of *NN* (3). Thus, Khomeini's two worst states (1 and 2), in Carter's view, were associated with the U.S.'s strategy of *I*.

Carter's imputation of these preferences to Khomeini turned out to be a major misperception of the strategic situation. Khomeini wanted the total Islamization of Iranian society and apostasy extirpated; the United States

[4] Game 50 is one of the 12 specific MG games, shown in figure 3.5, in which Magnanimity is an NME. In section 4.5 it was used as a model of bombing campaigns in Vietnam.

Game as misperceived by Carter (game 50)

		Khomeini Negotiate (N)		Khomeini Obstruct (O)	
Carter	Negotiate (N)	I Compromise (4,3) [4,3]	→	II Carter surrenders (2,4) [4,3]	←Dominant strategy
		↑		↓	
	Intervene militarily (I)	IV Khomeini surrenders (3,2) [4,3]/[2,4]	←	III Disaster (1,1) [2,4]/[4,3]	

Real game (game 5)

		Khomeini Negotiate (N)		Khomeini Obstruct (O)	
Carter	Negotiate (N)	I Carter succeeds (4,2) [2,4]	→	II Khomeini succeeds (2,4) [2,4]	←Dominant strategy
		↑		↓	
	Intervene militarily (I)	IV Carter adamant (3,1) [2,4]	←	III Khomeini adamant (1,3) [2,4] ↑ Dominant strategy	

Key: (x,y) = (payoff to Carter, payoff to Khomeini)
[x,y] = [payoff to Carter, payoff to Khomeini] in anticipation game (AG)
4 = best; 3 = next best; 2 = next worst; 1 = worst
Nash equilibria in original games and AG of game 50 underscored
NMEs circled
Arrows indicate direction of cycling

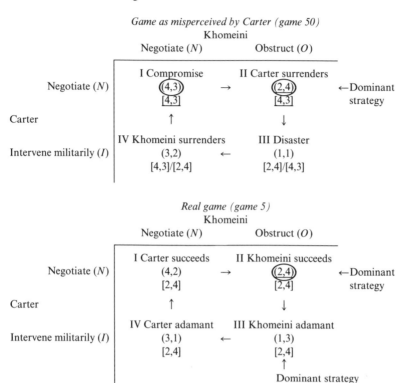

Figure 6.2 Iran hostage crisis (games 50 and 5)

was "a global Shah – a personification of evil" (quoted in Saunders, 1985, p. 102) that had to be cut off from any contact with Iran. Khomeini abjured his nation never to "compromise with any power ... [and] to topple from the position of power anyone in any position who is inclined to compromise with the East and West" (Sick, 1985a, p. 237).

If Iran's leaders should negotiate the release of the hostages, this would weaken their uncompromising position. Those who tried, including President Bani-Sadr and Foreign Minister Ghotbzadeh, lost in the power struggle. Bani-Sadr was forced to flee for his life to Paris, and Ghotbzadeh was arrested and later executed.

What Carter was unable to grasp was that Khomeini most preferred *O* (4 and 3), independent of what the United States did. Doubtless, Khomeini also preferred that the United States choose *N*, whatever his own strategy choice was, giving him the preferences shown in the bottom matrix of figure 6.2, which is game 5 and what I call the "real game."

Perhaps the most salient difference between Carter's game (game 50) and the real game (game 5) is that the former game contains two NMEs, (4,3) and (2,4), whereas the latter game contains only one, (2,4). In Carter's game, Carter's preferred solution of compromise at (4,3) can be reached wherever play commences, as is evident from its anticipation game (AG). In addition, (4,3) is not vulnerable to moving power: whichever player possesses it, (4,3) is the outcome, as I showed in section 4.5, making this kind of power irrelevant.

On the other hand, threat power is effective: Carter has a deterrent threat – threatening *I* to implement (4,3) – whereas Khomeini has a compellent threat – choosing *O* to implement (2,4). Thus, the possessor of threat power can implement its best outcome, as can the possessor of order power, starting from either (3,2) or (1,1). Insofar as Carter believed that he had the upper hand in game 50, therefore, he would see compromise is attainable by the exercise of any of the three kinds of power.

The prospects for compromise are very different in the real game (game 5). Not only is (2,4) the unique NME, favoring Khomeini, but Khomeini can induce it with a compellent threat. Carter does not have a threat strategy in this game; though game 5 is cyclic, moving power is ineffective (section 4.3); and order power is not defined because there are no indeterminate states. But because Carter thought he was playing game 50, in which the (4,3) NME can always be induced – even without the exercise of order power – if play starts at either state associated with *N*, Carter would have no reason not to choose *N*.

Adopting this strategy from the start turned out to be a blunder. However, in the game as he perceived it, Carter also had a deterrent threat (noted earlier) that could induce (4,3), which he pursued as well. He dispatched the aircraft carrier USS *Kitty Hawk* and its supporting battle group from the Pacific to the Arabian Sea. The carrier USS *Midway* and its battle group were already present in the area. Sick (1985b, p. 147) reported:

> With the arrival of *Kitty Hawk*, the United States had at its disposal the largest naval force to be assembled in the Indian Ocean since at least World War II and the most impressive array of firepower ever deployed to those waters.

But this threat, like those preceding it, did not lead to any change in Khomeini's strategy because of Carter's fateful underestimation of

Khomeini's willingness and ability to absorb economic, political, and military punishment in the pursuit of his revolutionary goals. In the real game, military intervention in Iran (*I*) leads to the (1,3) state when Khomeini chooses obstruct (*O*). Because of the possible execution of the hostages that this attack might provoke – the threat of which was "taken with deadly seriousness in Washington" (Sick, 1985b, p. 147) – I rank it as worst for the United States.

After negotiations faltered and then collapsed in April 1980, Carter was forced to move to his *I* strategy. If the rescue operation had succeeded and the hostages had been freed, the game would have been in state (3,1), because Khomeini could in that situation no longer use the hostages as a weapon and choose *O*.

The rescue's inglorious failure kept the situation in state (2,4) for another nine months. But the Iranian leadership had already concluded in August 1980, after the installation of an Islamic government consistent with Khomeini's theocratic vision, that the continued retention of the hostages was a net liability (Saunders, 1985, pp. 44–5). Further complicating Iran's position was the attack by Iraqi forces in September 1980. It was surely no accident that the day of Carter's departure from the White House on January 20, 1981 – 444 days after the capture of the hostages – they were set free.[5]

Although Carter's strategic acumen in this crisis can be questioned, it was less his rationality that was at fault than his misperception of Khomeini's preferences. Within a week of the embassy seizure, analysts in the State Department had reached the conclusion that

> diplomatic action had almost no prospect of being successful in liberating the hostages and that no economic or other U.S. pressure on the Iranian regime, including military action, was likely to be any more successful in securing their safe release. Consequently, they concluded, the detention of the hostages could continue for some months. (Sick, 1985a, p. 246)

In the first few months of the crisis, U.S. Secretary of State Cyrus Vance counseled that "we continue to exercise restraint" (Vance, 1983, p. 408). Privately, he vehemently opposed any military action; and after the military rescue operation failed, he resigned.

But others voiced different views, including secular politicians in Iran

[5] That the hostages were not released before the November 1980 presidential election, which clearly would have benefited Carter's bid for reelection, Sick (1991) attributes to a secret deal Iran made with Reagan supporters. But this allegation, at least as far as George Bush's involvement is concerned, is disputed by a bipartisan October Surprise Task Force of the U.S. House of Representatives chaired by Lee H. Hamilton (Lewis, 1992, 1993). The arguments for a secret deal (Sick) and against (Hamilton) are joined in "Last Word on the October Surprise?" (1993).

who claimed to speak for the implacable Khomeini. There was an abundance, rather than a dearth, of information, but the question, as always, was what was accurate. Carter, perhaps, should not be judged too harshly for misjudging the situation. In fact, even if he had foreseen the real game from the start, this analysis suggests that there was little that he could have done to move the state away from (2,4), given that the military power of the United States could not readily be translated into a credible threat.

Nevertheless, Carter's misperception gave him the hope that he could implement the compromise outcome in game 50, not only because it is an NME in that game but also because it can be induced via threat, moving, or order power. By contrast, the "Carter succeeds" outcome in game 5 is not an NME, and no kind of power can induce it.

This contrast between the two games is obscured by the classical theory, which shows (2,4), in which Carter surrenders, to be the unique Nash equilibrium, associated with Carter's dominant strategy of *N*, in each game. Thus, the classical theory makes Carter's actions inexplicable in terms of rational choice – whether he misperceived Khomeini's preferences or not – whereas TOM shows that Carter's actions were not ill-founded, given his misperception.

6.5 Deception in the Cuban missile crisis

Define a player's *deception strategy* to be a false announcement of its preferences to induce the other player to choose a strategy favorable to itself (i.e., the deceiver).[6] For deception to work, the deceived player must

(1) not know the deceiver's true preference ranking.

In fact, I interpret this assumption to mean that the deceived has no information about the deceiver.[7] Hence, the deceived has no basis for mistrusting the deceiver unless there is contrary evidence to indicate otherwise. I also assume the deceived does

(2) not have a dominant strategy.

Otherwise the deceived would always choose it, whatever the deceiver announced its own strategy to be. Obviously, a deception strategy, like a threat strategy, involves prior communication (PC).

Given PC and that conditions (1) and (2) are met, the deceiver, by pretending to have a dominant strategy, can induce the deceived to

[6] Game-theoretic models of deception, including some with applications, can be found in Brams (1977), Brams and Zagare (1977, 1981), Zagare (1979), Muzzio (1982), and Mor (1993). A useful compilation of material on deception, both in theory and practice, is Daniel and Herbig (1982).

[7] Although Carter misperceived Khomeini's preferences in the Iran hostage crisis, he was correct about Khomeini's most preferred state (4), namely *NO* (figure 6.2).

believe it will always be chosen. Anticipating this choice, the deceived will then be motivated to choose its strategy that leads to the better of the two states associated with the deceiver's (presumed) dominant strategy.

In the case of deception, I assume that play starts with PC and the choice of strategies by the players rather than commencing in a particular state. I shall return to this point later and discuss its consistency with TOM, but first I consider the possible use of deception in the Cuban missile crisis (section 5.4).

As the crisis heightened, the Soviets indicated an increasing predisposition to withdraw rather than maintain their missiles if the United States would not attack Cuba and pledge not to invade it in the future. In support of this shift in preferences, contrast two statements by Premier Khrushchev, the first in a letter to the British pacifist, Bertrand Russell, the second in a letter to President Kennedy:

> If the way to the aggressive policy of the American Government is not blocked, the people of the United States and other nations will have to pay with millions of lives for this policy. (Divine, 1971, p. 38)

> If assurances were given that the President of the United States would not participate in an attack on Cuba and the blockade lifted, then the question of the removal or destruction of the missile sites in Cuba would then be an entirely different question. (Divine, 1971, p. 47)

Finally, in an almost complete about-face, Khrushchev, in a second letter to Kennedy, all but reversed his original position and agreed to remove the missiles from Cuba, though he demanded a *quid pro quo* (which was ignored by Kennedy in his response, quoted in section 5.4):

> We agree to remove those weapons from Cuba which you regard as offensive weapons ... The United States, on its part, bearing in mind the anxiety and concern of the Soviet state, will evacuate its analogous weapons from Turkey. (Divine, 1971, p. 47)

Khrushchev, who had previously warned (in his first letter to Kennedy) that "if people do not show wisdom, then in the final analysis they will come to clash, like blind moles" (Divine, 1971, p. 47), seemed, over the course of the crisis, quite ready to soften his original position. This is not to say that his later statement misrepresented his true preferences – on the contrary, his language evoking the fear of nuclear war has the ring of truth to it. Whether he actually changed his preferences or simply retreated strategically from his earlier pronouncements, there was a perceptible shift from a noncooperative position (maintain the missiles regardless) to a conditionally cooperative position (withdraw the missiles if the United States would also cooperate).

Perhaps the most compelling explanation for Khrushchev's modification of his position is that there was, in Howard's (1971, pp. 148, 199–201) apt phrase, a "deterioration" in his original preferences in the face of their possibly apocalyptic consequences in the event of nuclear war. By interchanging, in effect, 3 and 4 in the Soviet ranking of states in the figure 5.4 representation of the crisis (game 30), Khrushchev made *W* appear dominant, thereby inducing the United States also to cooperate (choose *B*). The resulting (3,4) state is next best for the United States, best for the Soviet Union, and renders *BW* a Nash equilibrium in this putative game.[8]

Whether Khrushchev deceived Kennedy or actually changed his preferences, the effect is the same in inducing the compromise that was actually selected by both sides. Although there seems to be no evidence that conclusively establishes whether Khrushchev's shift was honest or deceptive, this question is not crucial to the analysis. True, I have developed the analysis in terms of rational deception strategies, but it could as well be interpreted in terms of a genuine change in Khrushchev's preferences, given that preferences are not considered immutable.

Could the United States have deceived the Soviets to induce (3,3) in game 30? The answer is no: if the United States had made *B* appear dominant, the Soviets would have chosen *M*, resulting in (1,4); if the United States had made *A* appear dominant, the Soviets would have chosen *W*, resulting in (2,2). Paradoxically, because the United States, as a deceiver, could not ensure an outcome better than its next worst (2) – whatever preference it announced – it was in its interest to be deceived (or at least induced) in order that (3,3) could be implemented.

More generally, in five of the 57 2×2 conflict games (29, 30, 31, 46, 47), at least one player can do better as the deceived than the deceiver, in terms of a player's comparative rankings of outcomes. Thus, it may be profitable not only that the deceived not know the preferences of the deceiver but also for the deceiver to know that the deceived does not know, and so on *ad infinitum* (Brams, 1977). For this set of five games, the strange notion that "ignorance is strength" – or "ignorance is bliss" – seems well founded.[9]

[8] This is game 21 in the Appendix, and (3,4) is also the unique NME in this game.

[9] These five games are a subset of the "deception-vulnerable" games, which comprise a total of 17 of the 78 2×2 strict ordinal games (22 percent); 21 games (27 percent) are "deception-proof," and 40 games (51 percent) are "deception-stable." Deception-vulnerable games are the only games in which deception is both possible – because both players do not have dominant strategies (unlike deception-proof games) – and profitable – because at least one player can induce a better outcome by announcing a false preference order (unlike deception-stable games). All the 2×2 games that fall into these mutually exclusive categories, as well as various subcategories, are given in Brams (1977).

Now consider the consequences for both sides if they had "played it safe" in the game 30 representation of the Cuban missile crisis by choosing their "security-level strategies." (A player's *security level* is the best payoff that it can ensure for itself, whatever contingency occurs, which in this case is the U.S.'s next-worst payoff of 2; a player's pure strategy associated with this payoff is its *security-level strategy*.) The choice of such a strategy to avoid its worst state (1) means the United States would choose A; if the Soviets also choose their security-level strategy (W), the resulting state is (2,2), which is Pareto-inferior to (3,3).

In section 5.4 I argued that not only is (3,3) the unique NME in game 30 but also that it can be induced by the moving power of either player. (In addition, it can be induced by a deterrent threat of the United States and a compellent threat of the Soviet Union.) Thus, (3,3) in game 30, even though this game is strongly cyclic (section 4.2) and has (2,2) as its security-level outcome, is quite attractive as a solution.

Nevertheless, I think Khrushchev had good reason to try to enhance its attractiveness to the United States by making it appear that he definitely would choose W because of its dominance. In seeming to interchange 3 and 4 in his preference ordering, he transformed game 30 into game 21, which entirely robs him of any incentive to depart from W.

This explanation for Khrushchev's about-face, grounded in the standard theory, makes a good deal of sense. But there is an alternative explanation, based on TOM, that seems to me equally plausible.

At the start of the crisis, the state was (1,4) in game 30, with the Soviets in the process of installing their missiles and the United States announcing that it would blockade further shipments. According to TOM, the migration from (1,4) to the NME of (3,3) in game 30 must proceed through the two other states, (4,1) and (2,2), as shown in figure 5.4. By contrast, in game 21, it is rational for the Soviets to move directly from (1,3) to the NME of (3,4).

Thus, the compromise state is, in a sense, more efficiently achieved in game 21 than game 30, according to TOM. This efficiency-based explanation of why Khrushchev sought to change the game offers an alternative way of viewing, within the TOM framework, the rationality of deception in this crisis.

I shall not analyze the "efficiency" of moves in games generally. But it is worth noting its relationship to the feasibility of moves discussed in section 1.6. There I argued that certain moves might be impossible, whereas here I argue that even if possible, players would presumably prefer to make fewer rather than more moves to reach an NME.

Changing the appearance of a game through deception may contribute to this greater efficiency. When it does, it is reasonable to suppose that players will use deception, especially if it can go undetected in games of incomplete information.[10]

6.6 Information revelation in *Hamlet*

There are other ways besides deception in which a player might manipulate information to try to gain an advantage. For example, instead of trying to deceive an opponent, a player might send a truthful signal in the hope of eliciting a response favorable to itself.

To illustrate this kind of strategy, consider the plot of William Shakespeare's most famous play, *Hamlet*.[11] There are actually several intertwined plots in *Hamlet*, but the main conflict is between the young Danish prince, Hamlet, and Claudius, Hamlet's uncle who becomes his stepfather after the death of Hamlet's father, the king of Denmark. Hamlet and Claudius stalk each other throughout the play, each seeking more information about the knowledge and motives of the other.

The intrigue begins when the ghost of Hamlet's father describes to Hamlet the true cause of his father's mysterious death: he was not stung by a venomous snake in an orchard, as supposed, but instead was murdered by Claudius, who poured poison into the king's ear as he slept. The apparition also reveals that Claudius was engaged in an adulterous affair with Hamlet's mother, Queen Gertrude, before the king's murder. The king's ghost asks Hamlet to "revenge his foul and most unnatural murder" (act I, scene v, line 30).[12]

Although Gertrude married Claudius with undue haste after the murder, the ghost tells Hamlet to ignore his mother's act, because she will suffer in her own way:

> Taint not thy mind, nor let thy soul contrive
> Against thy mother aught. Leave her to heaven,
> And to those thorns that in her bosom lodge
> To prick and sting her. (I, v, 92–5)

After the ghost vanishes, Hamlet is left doubtful of the ghost's appearance, even wondering whether it might be the devil.

Nonetheless, Hamlet vows to kill his uncle, now King Claudius, and

[10] The game-theoretic articles cited in note 6 distinguish between "tacit" deception, which is undetected, and "revealed" deception, which is detected. This distinction is illustrated by the 1954 Geneva negotiations over the future of the two Vietnams, which involved three players and are analyzed in Zagare (1979); see also Brams (1990, chapter 8).

[11] The subsequent analysis is adapted from the student papers of two former NYU undergraduates, Christina Brinson and Yasmin C. Jorge. For decision-theoretic exegeses of *Hamlet*, see Dalkey (1981) and Orbell (n.d.).

[12] Future references do not include the words "act, scene, line," which are those used in Shakespeare (1958).

thereby avenge his father's murder. The question then is how best to carry out this deed. Should he reveal his new-found but still uncertain knowledge of the treachery of his uncle (and now new stepfather), or should he keep secret his knowledge and dispatch him quickly?

Revelation, or even whispering his suspicions, would invite an attempt on Hamlet's own life by Claudius, who would then be able to surmise Hamlet's intentions. On the other hand, to hide his knowledge yet murder Claudius would make killing him seem an unjustifiable act as thoroughly dishonorable as Claudius' murder of Hamlet's father. Thus, Hamlet's two strategies are to reveal (R) his secret knowledge or not ($\bar{R}$).

Hamlet actually gives a subtle twist to his strategy of R. It involves hinting at his knowledge of the murder, in a way that would be apparent only to Claudius, to try to force Claudius to confirm, perhaps unconsciously, his evil deed. Hamlet does this by feigning madness, under the guise of an "antic disposition" (I, v, 197), to avoid being killed by a suspicious Claudius and to buy more time to gather incriminating evidence against him. The pretense confuses Polonius, the king's counselor: "Though this be madness, yet there is method in't" (II, ii, 222–3).

Claudius, too, tries to obtain more information about Hamlet's knowledge of the murder and Hamlet's intentions to avenge it, enlisting Hamlet's friends, Rosencrantz and Guildenstern, as spies. But Hamlet sees through them and engineers their execution. When Claudius eavesdrops on a conversation between Hamlet and Ophelia – Hamlet's lover (to whom he also feigns madness) – Hamlet admits that he is

> very proud, revengeful, ambitious; with more offenses at my beck than I have thoughts to put them in, imagination to give them shape, or time to act them in. (III, i, 134–7)

Claudius concludes that the young prince is not acting peculiarly because he is a harmless, lovesick fool. On the contrary, Claudius worries that "madness in great ones must not unwatched go" (III, i, 198–9).

As the evidence accumulates that Hamlet is a menace both to his life and his throne, Claudius must decide between killing Hamlet immediately (K), a risky strategy because of "the great love the general gender bear him" (IV, vii, 20), and not killing him ($\bar{K}$), which is also risky if Hamlet indeed intends to avenge his father's murder and slays him first.

In combination with Hamlet's strategies, the resulting game is shown in figure 6.3. The consequences of the players' strategy choices at the four states, moving clockwise from the upper left-hand state, are as follows:

I **Martyrdom for Hamlet: (3,2).** Hamlet reveals Claudius' treachery but is killed; Claudius is implicated in both murders and is punished.

		Claudius Kill (K)		Claudius Don't kill ($\bar{K}$)	
Hamlet	Reveal (R)	I Martyrdom for Hamlet (3,2) [3,2]	←	II Success for Hamlet (4,1) [3,2]	←Dominant strategy
				↑	
	Don't reveal ($\bar{R}$)	IV Failure for Hamlet (1,3) [3,2]		III Dishonor for Hamlet (2,4) [3,2]	

Key: (x,y) = (payoff to Hamlet, payoff to Claudius)
$[x,y]$ = [payoff to Hamlet, payoff to Claudius] in anticipation game (AG)
4 = best; 3 = next best; 2 = next worst; 1 = worst
Nash equilibrium in original game underscored
NMEs circled
Arrows indicate progression of states to NME

Figure 6.3 Hamlet–Claudius conflict (game 26)

II **Success for Hamlet: (4,1).** The murderer of Hamlet's father is uncovered, and Claudius is dethroned and killed.

III **Dishonor for Hamlet: (2,4).** Hamlet never reveals evidence to avenge his father's murder and is disgraced; Claudius is no longer threatened.

IV **Failure for Hamlet: (1,3).** Hamlet is dishonored for not revealing the murderer of his father and is killed as well; Claudius faces possible punishment for Hamlet's murder.

The game commences in state (2,4), where Hamlet hides his knowledge of the heinous crime and Claudius receives his best payoff. As TOM predicts, Hamlet (gingerly) switches to R. In an imaginative attempt to ferret out the truth, Hamlet puts on a skit for Claudius and Gertrude that reenacts the murder of his father and the clandestine affair between his mother and his uncle. Furthermore, Hamlet asks Horatio to observe Claudius' actions throughout the performance:

> Observe my uncle. If his occulted guilt
> Do not itself unkennel in one speech,
> It is a damned ghost that we have seen,
> And my imaginations are as foul
> As Vulcan's stithy [forge]. (III, ii, 81–5)

Horatio reports back that Claudius became pale during the performance and then, extremely agitated, walked out. Apparently guilt-stricken, Claudius was unable to bear the scene on stage. This confirms Hamlet's suspicions that the tale the ghost told was true, and that Claudius is the murderer.

Now the next move is Claudius' at (4,1). Even before the skit, his suspicions were aroused that Hamlet had uncovered the murder. But now his worst fears are confirmed, and he, like Hamlet, agonizes no more.

A series of events, too complex to relate here, unfolds. By the end of the play, Hamlet has revealed to Claudius his knowledge of the crime and slain him. But Hamlet himself succumbs to poison, as does his mother, a direct consequence of Claudius' actions.[13]

Thus, the players end up at the unique NME of (3,2) in game 26, which is also the Nash equilibrium associated with Hamlet's dominant strategy of *R*. What the Nash equilibrium fails to make apparent, however, is the rationality of the sequence of moves leading to it, from states III through II to I.[14]

Of course, even a move-by-move analysis does not begin to capture all the richness and subtlety of Shakespeare's great play. What TOM does demonstrate, however, is that the behavior of the principal antagonists is perfectly explicable in terms of rational moves, which is a view distinctly at odds with psychological interpretations of this play.

In particular, Hamlet had good reasons for acting bizarrely: first, to cover up his intentions so as not to get killed himself; second, to try to elicit the truth from Claudius about his perfidy. (Of course, Hamlet's behavior might also be interpreted as randomizing his choices to conceal his intentions, but the figure 6.3 game, interpreted cardinally, does not have a mixed-strategy Nash equilibrium.) Claudius was no less astute in trying to ascertain how much Hamlet knew about his fratricide.

Hamlet's failure to move decisively against Claudius, until the end, is perhaps less easy to fathom. His irresoluteness is often attributed to a tragic flaw in Hamlet's character – his melancholy nature or suicidal personality – which are viewed as psychological abnormalities. But is it so abnormal to be grief-stricken by the murder of one's father and lethargic after such an event?

Aside from inconsolable grief, Hamlet had cogent reasons for vacillating. He could not be sure about Claudius' complicity until after Horatio had witnessed Claudius' reaction to the skit. Also, he would have jeop-

13 In the play, Hamlet chooses *R* before Claudius chooses *K*, which suggests that Hamlet acted first, after which Claudius made his choice, in a 2×4 game, wherein Claudius' strategies depend on Hamlet's prior choices. But, in fact, Claudius contemplated killing Hamlet well before the skit, so the 2×2 game form is appropriate.

14 Interestingly enough, although moves from I to IV, and from IV back to III, are clearly infeasible (Hamlet cannot hide his knowledge of Claudius' complicity after letting it slip out, and Claudius cannot bring Hamlet back to life after killing him), these moves would not be made, even if they were feasible, because state I is the unique NME.

ardized his own ascension to the throne if he had acted rashly, before the evidence was incontrovertible.

Hamlet's scrupulous planning that led up to the entrapment of Claudius is, in my opinion, nearly flawless itself. But Claudius is no less adroit a game player. Like Hamlet, he could not act precipitously – by killing Hamlet – lest he be discovered as the perpetrator of two murders. In effect, each of the protagonists sought to peel away layers from the mystery surrounding the other.

Polonius, who, as I indicated earlier, had a dim view of what was going on, was also involved in the attempt:

> If circumstances lead me, I will find
> Where truth is hid, though it were hid indeed
> Within the centre. (II, ii, 171–3)

But the feckless Polonius, whom Hamlet kills before he can unravel the mystery, is no match for Hamlet.

I have made Hamlet's strategy choice one of revelation or nonrevelation, in which the purpose of revelation is to evoke a telltale reaction from Claudius. Though I posit Claudius to choose between killing or not killing Hamlet, bear in mind that he, like Hamlet, is vitally concerned with gathering and exploiting any information he can get his hands on before deciding whether to move against Hamlet.

Ultimately, of course, both players attempt to eliminate each other. Perhaps the most significant difference in their strategies is that Hamlet wants first to discover the truth, which requires revealing something about what he knows. Claudius wants to hide it but, at the same time, to discover what Hamlet knows.

Each player was slow to move against the other until he had assembled what he thought was overwhelming evidence. But once each had, their cat-and-mouse game turned diabolical and grim. Thus, Claudius tried to have Hamlet murdered in England, just as Hamlet plotted to kill Claudius when he returned to Denmark. In the end, tragedy befalls not only the antagonists but several other characters as well. Nevertheless, all the characters – including Hamlet – seem eminently rational in light of the informational constraints they faced.

6.7 The paradox of omniscience

Hamlet well illustrates a game in which information is at a premium. Throughout the play, each of the protagonists tries to learn more about the other's plans. As they perfect their information about each other, however, they draw closer and closer to the looming tragedy.

One wonders whether players, either in *Hamlet* or in real-life games, might sometimes do better by knowing less. Assume, for example, that Claudius is *omniscient*, by which I mean that he can predict Hamlet's strategy choice before he actually makes it. That would seem to give Claudius an advantage.

Suppose, however, that Hamlet is aware of Claudius' omniscience. Then Hamlet can ascertain, assuming he has complete information about Claudius' preferences as well as his own, that

- if he chooses R in game 26 (figure 6.3), Claudius will predict this choice and, as a best response, choose K, giving (3,2);
- if he chooses $\bar{R}$, Claudius will predict this choice and, as a best response, choose $\bar{K}$, giving (2,4).

Obviously, Hamlet prefers (3,2) and so will choose R, which, because he is omniscient, Claudius will predict and therefore choose K. Note that, in terms of the comparative rankings of the two players, Hamlet receives his next-best payoff and Claudius his next-worst. Relatively speaking, then, the omniscient player does worse than the nonomniscient one.

This comparison, however, can be challenged on two grounds. First, it involves an interpersonal comparison of utilities (or ranks in the ordinal case), and theorists have persuasively argued that such comparisons cannot be made. Who is to say that Hamlet's next-best payoff is better than Claudius' next-worst payoff unless there is a standard for making such a comparison, shared by both players? In fact, both players die in the end, in part because once Hamlet chooses R and Claudius' treachery is unmasked, Hamlet opts to kill his antagonist, just as Claudius chooses the same strategy against his opponent. Is it really possible to say who does better in death?[15]

The second challenge to the result that the omniscient player does worse than the nonomniscient one is that omniscience is less the issue than Hamlet's dominant strategy of R. In a game of complete information, Claudius hardly has to be omniscient to predict that Hamlet will choose R, which contains Hamlet's two best states (3 and 4). Thus, as long as information is complete, omniscience is beside the point: a nonomniscient player could as well predict that Hamlet will choose R, though not perhaps that Hamlet would try to cover up his knowledge by hiding it under the cloak of madness.

The omniscience of a player, coupled with an opponent's awareness of

[15] The fact that I label Hamlet a "martyr" at the (3,2) state is meant to reflect Hamlet's presumed willingness to sacrifice his own life to avenge that of his father's. But this ranking in no way implies that this is a better outcome for Hamlet than is death for Claudius, who ranks only one other state lower – not killing Hamlet after his revelation. The rankings by each player reflect their own assessment of each state relative to the other states, not how these assessments compare with those of the other player.

this omniscience, is more paradoxical in a game like Chicken (game 57), given in figure 5.2. Assume Column is omniscient. If Row chooses C, Column will predict this choice and choose $\bar{C}$ giving (2,4). On the other hand, if Row chooses $\bar{C}$, Column will predict this choice and choose C, giving (4,2). Because Row prefers (4,2), it will choose $\bar{C}$, forcing an omniscient Column to back down, receiving only its next-worst payoff (2). On the other hand, the nonomniscient Row obtains its best payoff (4).

Because the latter payoff cannot be improved on, whereas Column's most definitely can, there is more justification for saying that omniscience hurts a player (i.e., Column) in this game. In fact, I will say that there is a *paradox of omniscience* when it is better for both players to be nonomniscient than omniscient, assuming that the nonomniscient player is aware of its opponent's omniscience.[16]

This is not true in game 26 (figure 6.3). If Hamlet were omniscient rather than Claudius, and Claudius were aware of this fact, then the outcome would be (3,2) if Claudius chose K and (4,1) if he chose $\bar{K}$. Clearly, it is in Claudius' interest to choose K against an omniscient Hamlet and receive (3,2). Thus, neither player does better or worse by being omniscient – the outcome remains (3,2), which is both the unique Nash equilibrium and the NME in this game, whichever player is omniscient.

Besides Chicken, there are five 2×2 ordinal games (51–55) in which there is a paradox of omniscience (Brams, 1981, 1982a, 1982b, 1983). Neither player in any of these games has a dominant strategy, and each game has two Nash equilibria, which are also its NMEs. These are also games in which order power is effective, though order power is effective in four other games with indeterminate states as well.

To have omniscience in a game is equivalent to a player's moving second, giving it the opportunity to observe the prior choice of its opponent and respond to it. Omniscience simply eliminates the necessity of having to move second: it enables the omniscient player to anticipate its opponent's choice, without actually having observed it beforehand. But the nonomniscient player, aware that the omniscient player will rationally respond to its prediction, can capitalize on this fact in games vulnerable to the paradox of omniscience.

The resolution that TOM provides to the paradox of omniscience is that it enables players to depart from a state that the paradox induces

16 This test presumes that one player is omniscient and the other is not but is aware of its opponent's omniscience. One then compares the payoffs that both players receive when they are omniscient versus when they are not. If both players do better by being nonomniscient, then there is a paradox. Note that this definition avoids interpersonal comparisons, because it makes the standard not how well one does *vis-à-vis* an opponent but how well one does *vis-à-vis* oneself in the opposite role.

initially. In Chicken, for example, the paradox induces (2,4) or (4,2), depending on whether Row or Column is omniscient (see figure 5.2). Yet from each of these states the NME of (3,3) can be reached if Row or Column, respectively, has order power.

Unfortunately for the players in Prisoners' Dilemma (game 32), also shown in figure 5.2, there is no such resolution to another problem tied to the possession of omniscience. If either player is omniscient in PD, it is easy to show that the state induced is (2,2), from which there is no escape because it is an NME. Fortunately, PD is the only 2×2 game with a Pareto-inferior NME.[17]

I conclude that one player's omniscience, and the other player's awareness of it, is not always a blessing for the omniscient player. In fact, the omniscient player will prefer to be the nonomniscient player if (i) there is a paradox of omniscience, or (ii) only the other player can, by its omniscience, induce a Pareto-optimal outcome. On the other hand, with the exception of PD, players can always escape a Pareto-inferior initial state and move to a Pareto-superior NME, although there may be a conflict over which NME, if there is more than one, will be chosen.

Incomplete information, especially when compounded by misperceptions, is undoubtedly a greater problem for players in real-life games than omniscience, as I illustrated in the case of the Iran hostage crisis in section 6.4 and the Vietnam bombing campaigns in section 4.5. Nevertheless, there are games in which not only a lack of information, but also the deception of one player by another, can redound to the benefit of both, as I illustrated in the Cuban missile crisis.

My overview of some effects of different levels of information in games shows how information may influence both their play and the choice of an outcome. More information is not necessarily good for players, nor is being deceived necessarily bad, illustrating some of information's non-obvious consequences in games.

When information is incomplete about preferences, players still may make rational choices, as I illustrated in the generic Magnanimity Game, but it usually heightens their difficulties. When information is incomplete about the relative powers of players, as may have been the case in both the American Civil War and the several wars between Egypt and Israel between 1948 and 1973, there is likely to be a wrenching test of strength, which benefits neither player in the short run.

17 Although one player's omniscience induces a Pareto-inferior state in six other games, the other player's omniscience always leads to a Pareto-superior state. In these games, which are games 27–31 and 48, it is therefore in both players' interest for one player – and never the other – to be omniscient (Brams, 1982b).

In the longer run, however, the results of this test may instill a painful awareness of what is attainable by the use of force, inducing both sides to seek a peaceful resolution of their conflict. The resolutions achieved in the aforementioned cases studied here have been fairly robust, suggesting that hard lessons, caused in part by incomplete information, may not be easily undone.

7 Incomplete information in larger games: a model of negotiations

7.1 Introduction

In section 1.4 I suggested that the moves and countermoves of TOM might be thought of as modeling a bargaining process, in which states are either accepted (no move occurs) or rejected (move occurs); if the latter, the process moves on. Although this interpretation could be developed further, I take a different approach to modeling moves in games in this chapter, retaining some ideas of TOM but relinquishing others in order to develop a model of negotiation applicable to larger games.

Information in larger games, especially with more players, is likely to be harder to come by and utilize effectively. In addition, as I showed in chapter 6, the effects of information may be double-edged, sometimes benefiting players while at other times hurting them. An appreciation of this double-edged effect has ramifications for learning. In particular, because information is generally costly to acquire, it is useful to know when it may be worthwhile to do so.

For example, when new information dictates changing course, players may find it rational to *backtrack*, or reverse the direction of their moves, in games. Although backtracking is prohibited by the rules of TOM, it can be accommodated within TOM by making play of a game a function of the information players have about it at different times.

Assume, for example, that player *P*1 is playing a generic game like the Magnanimity Game (MG) and is not sure of its opponent's (*P*2's) type. *P*1 would, according to TOM, at least know that a specific move in MG may or may not be beneficial, depending on *P*2's type.

Once *P*1 has acquired this information in actually playing the game – based on *P*2's response to its move – it may change its notion of what game is being played, or narrow down the possibilities. Then *P*1 would effectively be playing a new game, starting at a new initial state, from which it may be rational to make a move that could be interpreted as backtracking in the old game.

Thereby, TOM effectively permits backtracking – but not by having an *ad hoc* rule that allows for this possibility. Instead, by allowing for the (1) acquisition of new information, which (2) may give rise to a new game, backtracking (in the old game) may become rational.[1] In other words, TOM permits backtracking if there is a new game, or if there is a new initial state in the old game from which it is rational to backtrack, which seems to me a more economical approach than encumbering the theory with additional rules.

In this chapter, I shall show how the framework of TOM, though not its specific rules, can be applied to games with more than two players, each of whom has a choice of more than two strategies in every state.[2] However, to simplify player calculations, I assume that the information that players have about other players is far more limited than in the 2×2 games analyzed so far. Thus, the theory presented here is not a generalization of TOM but an extension of its ideas about moves, and a curtailment of the information it assumes is available to players, in larger games.

More specifically, I posit that *the only information players have about each other's preferences is that which is revealed in the course of play*. This simplification makes it rational for players to follow a different set of rules from those given earlier for TOM, which I describe in section 7.3 and interpret in section 7.4.

At the start of play, I postulate that the players are at a point of disagreement: each prefers a different alternative and chooses different strategies to try to implement it. Then, as the game progresses, they acquire new information, which may or may not lead them to agreement in the end.

Unlike TOM, players do not select strategies on the basis of backward induction, simply because they do not have the information needed to make such calculations. Unable to look ahead for want of information, players instead base their choices at each stage on the current state of information that they have about the game.

I illustrate this analysis in the context of negotiations, in which a state is continually updated as the negotiations proceed.[3] Eventually a consensus

1 Actually, it is only perceptions about the old game that change. Still, this change means that players think the old game is different and, so, effectively, are playing a new game, not just playing a new round of the old game. Theories of repeated games (and supergames), on the other hand, postulate repetition of a single game. This assumption, in my opinion, is often untenable, because the very nature of a conflict is likely to change from round to round.

2 The remainder of this chapter, except for section 7.6, is based in part on Brams and Doherty (1993) and adapted with permission.

3 But I do not assume that the updating is based on Bayesian calculations, as is usual in the literature on noncooperative games of incomplete information (Rasmusen, 1989; Myerson, 1991; Fudenberg and Tirole, 1991; Binmore, 1992). Such calculations require more information than is assumed by TOM – in particular, probabilistic information on

is reached, which may be either agreement on an alternative or impasse (i.e., an agreement to disagree).

To determine the kind of consensus that is achieved, I assume a voting rule. This need not be fixed but can be modified to fit the situation being modeled. (For example, consensus may be based on unanimity rather than majority rule, which I postulate here.) Like TOM, players can have different powers, but these differences are tied to their different abilities to affect the outcome, based on their size, rather than on their being able to continue moving, choose the order of play, or threaten their opponents.

I have chosen to model negotiations, rather than more general conflict processes, because I wish to focus on the conditions that lead to *conflict resolution* in multiplayer, multi-alternative situations. This is not a severe limitation on the study of n-person conflicts, because almost all such conflicts involve negotiations – typically, with more than two alternatives on the table – aimed at achieving a settlement.

The model of negotiations I shall present, though it differs drastically from TOM as developed so far, is very much in its spirit. First and foremost, it allows for sequences of moves and, therefore, has the dynamic character of TOM. Second, players may have different abilities, or powers, to affect the outcome. However, because the setting is more complex – with more than two players, each having more than two choices – I restrict the players' information in order to make the analysis tractable.

Such a restriction is probably realistic in most negotiations. Of course, it significantly changes the nature of rational calculations, which I make explicit with a

> **Revelation assumption (RA):** In a game of incomplete information, players have no information about the preferences of the other players, except as they are revealed in the course of negotiations.

The principal question I explore in this chapter is, given RA, the rationality of being intransigent in negotiations. It is often alleged that being tough at the bargaining table pays dividends to a negotiator if, in the end, an agreement emerges (presumably favorable to the tough negotiator). But, at the same time, toughness on the part of all parties – an unwillingness to compromise – may sabotage the possibility of any agreement, which may hurt everybody.

states, and cardinal utilities rather than ordinal payoffs for the players – which Skyrms (1990), among others, assumes in his "dynamics of rational deliberation." For another game-theoretic model for analyzing the transformation of conflict into cooperation through negotiation, see Brams (1992).

Trade-offs between holding out, on the one hand, and giving up too much, on the other, have been extensively analyzed in two-person negotiation games (Raiffa, 1982; Brams, 1990; Young, 1991). Much less is known, however, about the effects of intransigence, or the lack thereof, in *n*-person games (but see Sebenius, 1984), in part because of the complications that the inclusion of new players with new interests brings to the analysis.

Although I restrict the subsequent analysis to three-person conflicts, the players may have different weights and, therefore, different effects on the outcome. Because the weightier players may encompass several individual players who decide to coordinate their actions or form a coalition, the analysis is also applicable to larger games, in which the players may be thought of as factions of possibly different size.

To motivate the analysis, I discuss an example in section 7.2, with a tie-breaker, in which the players progressively invoke fallback positions to try to prevent their worst states from becoming the outcome. They aim, instead, at achieving at least a next-best outcome in a game of incomplete information.

In section 7.3 I introduce and illustrate the new rules of play without a tie-breaker. These rules allow for the possibility of impasse, which some players may prefer to certain alternatives. I discuss the effects of impasse, and interpret the rules with more examples, in section 7.4.

The intransigence of players, or their unwillingness to retreat to fallback positions, generally works to their advantage in effecting preferred outcomes, as shown in section 7.5. Although greater size also helps a player, intransigence is a potent force by itself, given that all preference orders are equally likely. In practice this assumption is surely not accurate, so the value of intransigence needs to be considered in a specific context.

Toward this end, I analyze recent negotiations on world trade in section 7.6, focusing on the three-person game played among the United States, the European Community (EC), and Japan under the auspices of GATT (General Agreement on Tariffs and Trade). I consider each player's position on two major issues and postulate preferences, based on the players' presumed primary and secondary goals. The analysis indicates that the key to a settlement (as of this writing in April 1993) will be how unacceptable each of the players considers the present deadlock.

Emphatically, the negotiation model developed here does not assume that all players are interested in "getting to yes" (Fisher and Ury, 1983). On the contrary, it allows for deadlock if fallback positions, beyond a certain point, are less desirable than accepting an impasse. Such behavior, as I demonstrate, may be eminently rational, which is not always recognized by those who deplore the lack of agreement in multilateral negotiations like those of the GATT.

7.2 A negotiation game with a tie-breaker

I begin by illustrating a model of sequential negotiation, based on incomplete information, and then indicate its relationship to a model of "sophisticated voting," based on complete information.[4] Although the assumption of incomplete information sometimes complicates matters, as I illustrated in section 4.5 and chapter 6, in this case it facilitates the calculations of players in a plausible manner.

Common to both the complete-information and incomplete-information models is a player who can break a tie if there is an impasse. Hence, there can never be an indeterminate outcome. The possibility of indeterminacy will be incorporated into the more general analysis of sequential negotiation games later, in which a deadlock can arise and not necessarily be broken.

Assume three players, *A*, *B*, and *C*, are trying to negotiate an agreement. They have the following preferences for alternatives *a*, *b*, and *c*:

A: *abc* (*a* preferred to *b*, *b* preferred to *c*)
B: *bca*
C: *cab*

These preferences result in a *paradox of voting*: $a>b>c>a$, where "$>$" indicates "is preferred by a majority to."[5]

Assume that each of the players knows only its own preferences but does not know the other two players' preferences (except as they are revealed in the course of play). The negotiation process, I assume, unfolds as follows:

1 The players begin by announcing their first choices. This is not an unreasonable assumption; not knowing the other players' preferences, each has nothing to lose by revealing its own most preferred alternative. In figure 7.1, I show this revelation at stage 1 by placing a vertical bar

[4] This section is based in part on Brams (1991). *Sophisticated voting* involves the successive elimination of dominated strategies in a game of complete information, given that other voters do likewise (Farquharson, 1971).

[5] Thus, *a* is preferred to *b* by *A* and *C*, *b* is preferred to *c* by *A* and *B*, and *c* is preferred to *a* by *B* and *C*. The fact that every alternative can be defeated by some other alternative creates *cyclical majorities*, or a cycle of social preferences such that no alternative is the social choice because a majority prefers it to another alternative. Although the lack of any social consensus is particularly acute when there is a paradox of voting, the other "genuine conflicts" I shall analyze later all involve a disagreement over first choices among the three players.

Player	Preference	Stage 1	Stage 2	Stage 3	Stage 4
A	*abc*	*a*|*bc*	*a*|*bc*	*ab*|*c*	*ab*|*c*
B	*bca*	*b*|*ca*	*bc*|*a*	*bc*|*a*	*bc*|*a*
C	*cab*	*c*|*ab*	*c*|*ab*	*c*|*ab*	*ca*|*b*
State		*a*	*c*	*b*	*a*

Note: A is assumed to have a tie-breaking vote, so in the event of a two-way or three-way tie, the tied alternative that *A* supports will win (*a* in stage 1, *b* in stage 3, and *a* in stage 4).

Figure 7.1 Sequential negotiation model with tie-breaking vote

between the players' first and second choices, indicating that each supports only its first choice (to the left of the line).

In the event of a three-way tie among *a*, *b*, and *c* (each alternative receives, in effect, one vote), assume A is the "chair"; its first choice, *a*, will prevail by virtue of *A*'s having a tie-breaking vote. This assumption of a tie-breaking vote – or, equivalently, an edge in negotiations should an impasse develop – precludes an indeterminate outcome. Thus, each alternative obtains one vote in stage 1, but *A*'s tie-breaking vote makes *a* the winner.

2 I call this (temporary) winning outcome a *state*, which is consistent with the usage of TOM (section 1.3). I assume that it induces *B* – for whom *a* is its worst choice – to fall back to its second choice, supporting *c* as well as *b*. Alternative *c* will then have a majority of two out of three supporters (players *B* and *C*), which I assume is sufficient to win, so I indicate *c* as the winning state at stage 2.

3 But now, because *c* is *A*'s last choice, it is *A* who must worry that its worst alternative will be selected. Consequently, it is reasonable to suppose that it will be *A*, next, that will expand its support to include *b*. The resulting two-way tie (two votes each) between *b* and *c* at stage 3 will be broken in favor of *b* because *A*, the tie-breaker, supports *b* over *c*. Hence, *b* becomes the winning state at stage 3.

4 Now, however, it will be *C*'s turn to throw its support to *a* (as well as *c*), lest *b* win at stage 3. At stage 4, then, *a* will win because of *A*'s tie-breaking vote, and there is now nothing either *B* or *C* can do to alter this outcome. It is in equilibrium, particularly to the chagrin of *B*, for whom *a*

is its worst choice. Even if *B* withdraws its support from *c*, however, *a* will still prevail.

The simple logic of this model, which makes the choices of the players endogenous – depending on what new information was divulged in the previous round – seems quite compelling in negotiations: each player is motivated, in turn, to expand its support from first to second choices at different stages to try to prevent its worst state from becoming the outcome. By stage 4, all players have revealed their top two choices and, therefore, a complete preference ordering over the three alternatives. As one would expect, the alternative that wins is the tie-breaker's first choice.

This game of incomplete information that I have informally described (rules will be specified later) has a counterpart as a game of complete information, in which each player knows everybody's complete preference ordering from the start. It takes much more complicated and less realistic calculations in the complete-information case to show that the outcome is exactly the same when sophisticated voters use "approval voting."[6] That is, *a* wins, with each voter voting for its top two alternatives.

But in a game of complete information, the plausibility of the sophisticated-voting model is unduly stretched. To determine that *a* wins under this model, players must first ascertain that *B* has a dominant strategy of voting for both *b* and *c* in a three-person game of complete information. Knowing that *B* will never vote for *b* alone (i.e., because this strategy is dominated), they next must determine that *A* has a dominant strategy of voting for both *a* and *b* in the reduced game (with *B*'s dominated strategy eliminated). Knowing in turn that *A*'s strategy of voting for *a* alone is dominated – reducing the game still further – they then must determine that *C* has a dominant strategy of voting for both *c* and *a*. Thereby the successive elimination of dominated strategies by the players gives *a* as the sophisticated outcome; the calculations for plurality as well as approval voting are given in Brams, Felsenthal, and Maoz (1986, 1988).

Unlike the sophisticated-voting model, the negotiation model does not assume major information-processing and calculational abilities on the part of the players but, rather, successive movements by the players to their fallback positions (i.e., second choices in the example) as the possibility of ending up with their worst choices looms. It seems to mimic how players might actually think as they negotiate over alternatives. Indeed,

[6] Under *approval voting*, voters can vote for as many alternatives as they like (Brams and Fishburn, 1983), mirroring the ability of players in negotiations to take fallback as well as most-preferred positions. Under *plurality voting*, by comparison, voters can vote for only a single alternative, which in negotiations would not permit them to support a fallback position as well as a first choice.

an *n*-person negotiation model should not be overly complex or esoteric if it is to capture how real-life players weigh their options and choose one from among them.

Although the negotiation model duplicates the results of the sophisticated-voting model in the preceding example, it does not involve the kind of optimality calculations usually made in game theory or TOM. Rather, I prescribe the choices that players make – instead of deriving them from players' preferences – and then trace out their step-by-step consequences. These consequences, which are by no means obvious, are based on physical moves that players might plausibly make over time rather than mental calculations of optimal behavior, which players cannot make because of a lack of information about the preferences of their opponents (which were assumed in earlier TOM calculations).

7.3 The tie-breaker removed

In going from stage 1 to stage 4, the players' progressive expansion of their support for alternatives is fueled by the fact that, at each stage prior to stage 4, there is a (temporary) winner that is the worst state of one of the players. Moreover, this player, by giving its support to a lower-ranked alternative, can induce a better state for itself, at least temporarily.

By stage 4, however, *B*, who ranks the stage 4 winner (i.e., *a*) last, is helpless to induce a better outcome. The only alternative that *B* does not support is *a*, its worst choice, which wins at stage 4 without *B*'s support.

Now assume that no player has a tie-breaking vote, making the game a symmetrical one for the players. Thus, at stage 1, there would be a three-way tie among *a*, *b*, and *c*. I call the lack of a majority consensus for a single alternative impasse (*i*).

Assume that a player not only is able to rank the alternatives *a*, *b*, and *c* but also has a preference for *i*. Thus, each player inserts *i* as an alternative in its preference order. For example, if *A* prefers *a* to *b*, *b* to *i*, and *i* to *c*, its ranking is *abic*.

Now consider the earlier example, in which *i* is the winner at stage 1 (no single alternative is supported by a majority) if *A* does not have a tie-breaking vote. The fact that *A*'s preference order is *abic* means that it would choose *b* next; if the other players have analogous preferences for *i*, they would all support their second choices as well as their first, moving them to stage 4 in figure 7.1. Then each player would "choose" *i* rather than support its worst choice, making *i* the outcome.

To formalize this process, assume that a player's support for the three "regular" alternatives – *a*, *b*, and *c* – as well as *i* unfolds according to the following rules of play and rationality rules:

1 Each player supports its first choice, which is assumed not to be *i*, at the start of the game (stage 1 earlier).
2 If the resulting state, which may be *i* (if there is a tie), is the same or less preferred than the player's second choice, then the player supports its second choice.
3 At any subsequent stage, if the resulting state is the same or less preferred than the player's next-lower choice, then the player supports this choice.
4 Play terminates when the state for all players is not worse than their next-lower choice (if there is one), which then becomes the outcome.

In the earlier example, in which all players ranked *i* just above their worst alternative, they would begin by supporting their first choices, which would create a three-way tie and therefore lead to *i* (stage 1 earlier). Because *i* is worse than their second choices, they would next support these choices as well, which would create another three-way tie and again lead to *i* (stage 4 earlier). Because this state in which *i* wins is the same as the players' next-lower choice (i.e., *i*), the players would all support *i* and play would terminate, making *i* the outcome.

Interestingly enough, if *A* ranked *i* second, making its ranking *aibc*, whereas the rankings of *B* and *C* remained *bcia* and *caib*, the outcome would be the same. As before, it would be *i* at the end of the first stage; in the second stage, *B* and *C* would lower their support to *c* and *a*, respectively – creating a tie between these two alternatives that gives *i* – and *A* would also support *i*. In the third stage, *B* and *C* would lower their support once again (to *i*), at which point play would terminate, making *i* the outcome. On the other hand, if *A* ranked *i* last, making its ranking *abci*, and *B* and *C* were as before, the outcome would be *c*, because *A* would give its support to *c* (before *i*) in the final stage, making *c* the winner (it would have support from all three players).

With these examples in mind, I next examine more systematically how the placement of *i* affects outcomes under the four rules of play. For the paradox-of-voting preference scales of the players in the earlier example (I shall include other possible preference scales in the later analysis), there are several different ways in which *i* can be incorporated into the players' preference scales to yield the same outcome. I list these below and indicate whether the player(s) who rank *i* higher (i.e., are more intransigent in the sense that they find impasse more acceptable) do better than those who rank *i* lower:[7]

[7] If the names of players *A*, *B*, and *C* are changed in the listing below, appropriate changes must be made in the outcomes. (In case 4, for example, if *A* rather than *C* is the player who ranks *i* second, *a* rather than *c* is the outcome.) Only case 5 gives qualitatively different

1 *A, B, and C rank i the same.* Whether the players all put *i* in second, third, or fourth place, there will always be a three-way tie among *a*, *b*, and *c* at the point at which the players cease their support of lower-ranked alternatives. The outcome is therefore *i*, and the effect of different rankings of *i* is moot because they are the same for all players.

2 *A and B rank i the same (e.g., fourth); C ranks i one rank higher (e.g., third).* *a* and *c* tie for the most support (e.g., from three players), and *b* receives less (e.g., from two). The outcome is therefore *i*, so *C*'s higher ranking of *i* does not help *c*.

3 *A and B rank i the same (e.g., third); C ranks i one rank lower (e.g., fourth).* *b* receives more support (e.g., from three players) than does *a* or *c* (e.g., from two players). The outcome is therefore *b*, so *A*'s and *B*'s higher ranking of *i* may help one player (*B* in this case).

4 *A and B rank i fourth; C ranks i second.* *c* receives support from three players and *a* and *b* from two players each. The outcome is therefore *c*, so *C*'s higher ranking of *i* does help *c*.

5a *A ranks i second; B ranks i third; C ranks i fourth.* Each of *a*, *b*, and *c* receives support from two players, creating a three-way tie. The outcome is therefore *i*, so *A*'s and *B*'s higher ranking of *i* does not help *a* or *b*.

5b *A ranks i second; C ranks i third; B ranks i fourth.* *a* receives support from three players, *c* from two players, and *b* from one player. The outcome is therefore *a*, so *A*'s higher ranking of *i* does help *a*.

6 *A and B rank i second; C ranks i fourth.* Each of *a*, *b*, and *c* receives support from two players, creating a three-way tie. The outcome is therefore *i*, so *A*'s and *B*'s higher ranking of *i* does not help *a* or *b*.

7 *A and B rank i second; C ranks i third.* Each of *a* and *c* receives support from two players and *b* from one player, creating a two-way tie. The outcome is therefore *i*, so *A*'s and *B*'s higher ranking of *i* does not help *a* or *b*.

To summarize, in the seven of eight cases in which players rank *i* differently (the rankings are the same in case 1), a player who ranks *i* higher

- is not helped in four cases (2, 5a, 6, and 7) – *i* is the outcome in each
- may be helped in one case (3)
- is helped in two cases (4 and 5b).

I hasten to add that this breakdown of cases does not presume that each case will occur with the same relative frequency. Consider cases 2 and 4,

outcomes if the names are changed – *i* wins in 5a, but *a* wins in 5b – which is why I distinguish these two cases.

but now assume that it does not matter which player ranks *i* in a certain way – it could be any of the three players. Thus in case 2, there are three ways in which two players can rank *i* fourth and the remaining player ranks *i* third, and three ways in which two players can rank *i* third and the remaining player ranks *i* second, making for a total of six ways. By comparison, there are only three ways in which case 4 can occur.

Yet, this is not to say that case 2 will occur twice as frequently as case 4. Empirically, the opposite might be the case if there tends to be one player who is much more intransigent than the others (by ranking *i* two, rather than one, level higher than the other two players). What the incorporation of *i* into the players' preferences establishes is that being less accommodating than other players sometimes helps a player achieve a preferred outcome.

Intransigence never hurts a player. Even when a player's higher ranking of *i* does not help it, as in cases 2, 5a, 6, and 7, the intransigent player may still benefit. The reason is that its greater favorableness toward *i* over supporting lower alternatives implies that it has a higher preference for *i* than player(s) who rank *i* lower. Hence, even if it does not win with its most preferred "regular" alternative, the intransigent player is comparatively better off in realizing its aims when *i* is the outcome.

7.4 Interpretation of the rules

So far I have shown that when the players have paradox-of-voting preferences, the more intransigent player(s) generally benefit by ranking *i* higher than the other player(s). But these preferences represent only one set of rankings that players may have. In section 7.5, I shall systematically analyze the effects of intransigence as well as player size for all possible rankings by three players, but next consider how rules 1–4 apply when the preferences of the players are not necessarily paradoxical.

I confine the analysis here and later to *genuine conflicts*, in which the three players all have different first choices: *A* most prefers *a*, *B* most prefers *b*, and *C* most prefers *c*. Thus, there is disagreement at the start of negotiations – or, based on rule 1, there is impasse initially.

Rule 2 is next invoked, with each player supporting its second choice. This second choice may be one of the two remaining alternatives (e.g., either *b* or *c* if *A* is the player) or *i*. Because there are $3! = 6$ ways of ordering these three lower-ranked options for each of the three players, there are $6^3 = 216$ possible orderings of them for all three players.

Applying rules 2–4 to the selection of an outcome, I first ascertain in how many of these orderings there is impasse. As an illustration of how such an outcome is arrived at, consider the following orderings of *A*, *B*, and *C* (read from left to right):

A:	*a* | *ibc*	*ai* | *bc*	*ai* | *bc*
B:	*b* | *aic*	*ba* | *ic*	*bai* | *c*
C:	*c* | *bia*	*cb* | *ia*	*cbi* | *a*
State:	*i*	*i*	*i*

As in the earlier example, the vertical lines in the orderings just to the right of *A*, *B*, and *C* indicate that the players start off by supporting only their first choices (rule 1). Because, by assumption, the first preferences of the three players all differ at this stage, *i* is the initial state.

In the second stage, each player supports its second choice as well (rule 2). At this stage, both *a* and *b* are supported by two players (*c* is supported by only one player), so *i* is the second state because of a two-way tie. (*A*'s support of *i* at this stage simply reinforces this choice.) In the third stage, *B* and *C* support their third choices (rule 3), giving unanimous support to *i*,[8] which terminates play since this state is not worse than the next-lower choices of each player (*b*, *c*, and *a* of *A*, *B*, and *C*, respectively) and therefore becomes the outcome.

But note that *b* is the so-called *Condorcet alternative* in this example: majorities prefer it not only to *i* but also to *a* and *c*. Nevertheless, *i* is the social choice in the negotiation model, illustrating that, as in elections, the Condorcet alternative does not always win (Brams, 1985a, chapter 4).

To determine whether *i* is the outcome, however, I do not compare player support for it with that for *a*, *b*, and *c*. The following example illustrates this point:

A:	*a* | *ibc*	*ai* | *bc*	*aib* | *c*
B:	*b* | *ica*	*bi* | *ca*	*bi* | *ca*
C:	*c* | *bia*	*cb* | *ia*	*cb* | *ia*
State:	*i*	*b*	*b*

At the second stage, both *b* and *i* are supported by two players, but *b* nevertheless is the state because it is the only regular alternative with majority support. (This becomes unanimous support at stage 3 and makes *b* the outcome.) Recall that *i* is the state only if there is no single regular alternative with the most support (i.e., because of a tie).

If there is unanimous support for a regular alternative at any stage, as there is in the next example, this alternative defeats one supported by a simple majority:

[8] Although *i* has more support than any regular alternative at this stage, this is not why it wins, as the next example illustrates.

<table>
<tr><td>A:</td><td>a | bci</td><td>ab | ci</td></tr>
<tr><td>B:</td><td>b | cai</td><td>bc | ai</td></tr>
<tr><td>C:</td><td>c | bia</td><td>cb | ia</td></tr>
<tr><td>State:</td><td>i</td><td>b</td></tr>
</table>

At stage 2, *b* is the alternative supported by all three players, whereas *c* is supported by two players and *a* by only one player.

None of the examples so far illustrates the situation in which it is rational for a player to give its support to alternatives it ranks below *i*. Yet rules 1–4 allow for this possibility, as the following example illustrates:

<table>
<tr><td>A:</td><td>a | ibc</td><td>ai | bc</td><td>ai | bc</td><td>ai | bc</td><td>ai | bc</td></tr>
<tr><td>B:</td><td>b | ica</td><td>bi | ca</td><td>bic | a</td><td>bic | a</td><td>bic | a</td></tr>
<tr><td>C:</td><td>c | abi</td><td>ca | bi</td><td>ca | bi</td><td>cab | i</td><td>cabi |</td></tr>
<tr><td>State:</td><td>i</td><td>a</td><td>i</td><td>i</td><td>i</td></tr>
</table>

Because *B* ranks *c* higher than *a* (the state at the second stage), at the third stage it will lower its support to include *c*. However, its support of *c* creates a two-way tie between *a* and *c*, resulting in impasse at the third stage, which *B* actually prefers to *c*. But now *C*, which prefers *b* to *i*, will lower its support to *b* at the fourth stage; at the fifth stage, *C* will also support *i* because it is the state at the previous stage. Play terminates at the fifth stage because *i* is not worse than each player's next-lower alternative (if there is one).

But note that both *A* and *C* prefer *a* to *i*. However, because there is a paradox of voting – a majority (*B* and *C*) prefers *c* to *a*, a majority (*A* and *B*) prefers *b* to *c*, and a majority (*A* and *C*) prefers *a* to *b* – one cannot single out one alternative as socially preferred.[9]

Observe that alternative *a* (rather than *i*) would win if *C* misrepresented its preference to be *acbi* rather than *cabi*, because *a* would have a majority at the first stage. But this misrepresentation violates rule 1 of the model. Moreover, if players have no information about the preferences of the other players until they are revealed, they have no basis for devising an "optimal" deception strategy (section 6.4).[10]

7.5 Intransigence versus size: which is more helpful?

Now consider, for the 216 different ways in which three players may order the three options below their first choices, in how many they encounter impasse and in how many one of *a*, *b*, or *c* wins. I have done this analysis

9 This example of the paradox is case 6 in section 7.3, in which *i* is the outcome.

10 For different game-theoretic models of deception, see Brams (1978), Brams and Zagare (1977, 1981), and Mor (1993).

for four different *weight configurations*, whereby *A*, *B*, and *C* may have different effects on the outcomes because of their different sizes.

These configurations are defined by weighted votes that I assign to the players. In all configurations, a simple majority of votes is assumed to be sufficient for an alternative to win – provided that it does not tie with another, in which case there is impasse:

1 **All players equal** (*A*, *B*, and *C* have 1 vote each): *a*, *b*, and *c* win in 44 combinations each (total: 132); *i* wins in 84.
2 **One large player and two small players** (*A* has 2 votes, *B* and *C* have 1 each, giving *A* a veto[11]): *a* wins in 76 combinations, *b* and *c* in 41 each (total: 158); *i* wins in 58.
3 **Two large players and one small player** (*A* and *B* have 2 votes each, *C* has 1 vote): *a* and *b* win in 72 combinations each, *c* in 34 (total: 178); *i* wins in 38.
4 **One large, one medium, and one small player** (*A* has 3 votes, *B* has 2 votes, and *C* has 1 vote, giving *A* a veto): *a* wins in 98 combinations, *b* in 60, and *c* in 34 (total: 192); *i* wins in 24.

Even though *A* has a veto in both configurations 2 and 4, ties – and therefore outcome *i* – are less common in configuration 4 because of the three different weights of the players. For example, if *A* supports *a* and *c*, *B* supports *a* and *b*, and *C* supports *b* and *c*, there is a tie between *a* and *c* in configuration 2 (with 3 votes each; *b* has 2 votes), whereas *a* wins in configuration 4 (with 5 votes; *c* has 4 votes and *b* has 3 votes).

Define the *determinacy ratio* of configuration *j* as the following quotient:

$$\text{DR}_j = (\text{total no. of wins of } a,\ b, \text{ and } c)/216,$$

or the ratio of the total number of wins of the three regular alternatives to the total number of wins of both the regular alternatives and impasse in all the preference orderings. For each of the four configurations, $\text{DR}_1 = 0.61$, $\text{DR}_2 = 0.73$, $\text{DR}_3 = 0.82$, and $\text{DR}_4 = 0.89$. Clearly, impasse is less likely as players of different size, who can affect the outcome differentially, negotiate according to the model.

Define the *power ratio* of a configuration *j* as the following quotient:

$$\text{PR}_j = (\text{no. of wins of large player})/(\text{no. of wins of small player}),$$

or the ratio of the number of wins of a large player to the number of wins of a small player (ignoring the medium player in configuration 4). For each of the four configurations, $\text{PR}_1 = 1$, $\text{PR}_2 = 1.85$, $\text{PR}_3 = 2.12$, and

11 By *veto* I mean the ability to block the action of a coalition of the other two players by creating a tie. Put another way, without the vetoer's votes, a coalition of the other players

Configuration	Player type	Total wins	Values (i_2, i_3, i_4)	Proportions
1 All players equal	Same	44	(26,10,8)	(.59,.23,.18)
2 One large and two small	Large	76	(56,12,8)	(.74,.16,.11)
	Small	41	(19,12,10)	(.46,.29,.24)
3 Two large and two small	Large	72	(36,18,18)	(.50,.25,.25)
	Small	34	(18,8,8)	(.53,.24,.24)
4 Large, medium, and small	Large	98	(58,20,20)	(.59,.20,.20)
	Med.	60	(28,16,16)	(.47,.27,.27)
	Small	34	(14,10,10)	(.41,.29,.29)

Figure 7.2 Values (i_2, i_3, i_4) and proportions for each weight configuration

$PR_4 = 2.88$, which demonstrates the greater relative ability of the large player to achieve its most preferred outcome going from configuration 1 (no large player) to configuration 4.

In configuration 4, the ratio of the wins of the large player to the wins of the medium player is 1.63, showing that the large player has a win advantage over the medium player greater than their weight ratio of $3/2 = 1.5$. However, the $2/1 = 2.0$ weight ratio of the medium player to the small player is not matched by the win advantage of 1.76 that the medium player enjoys over the small player.

I conclude that the greater weight of large players enables them more often to win than smaller players. However, the translation is only approximate: A player's ability to implement preferred outcomes is only roughly in proportion to its weight.

Consider next the effect of a player's *intransigence*, measured by how high in its preference order it ranks i. The results for each configuration are given in figure 7.2, where each triple (i_2, i_3, i_4) indicates the number of wins of each player when it ranks i second, third, or fourth in its preference order. I have also given the proportions of wins for each ranking to facilitate comparisons within and between configurations.[12]

The large player does spectacularly better in configuration 2 when it ranks i second rather than third, winning almost five times as often ($56/12 = 4.67$). It wins almost three times as often in configuration 4

would not have a majority, so the vetoer's votes are necessary, though not sufficient, for a coalition to win.

[12] The values of i_3 and i_4 for the small players in configurations 2 and 4 are averages. (In configuration 2, 10 is an average of 12 and 8; in configuration 4, 10 is an average of 9 and 11.) These averages iron out the (small) differences between the wins *B* versus *C* enjoys when *A* – instead of *B* or *C* – is the large player.

(58/20 = 2.90), and twice as often in configuration 3 (36/18 = 2.00), when it is similarly intransigent. Surprisingly, it makes no difference, on the average, if the large player ranks *i* third or fourth in configurations 3 or 4 – it does equally badly – whereas in configurations 1 and 2 ranking *i* third is more helpful than ranking *i* fourth. In sum, intransigence never hurts and often benefits players.

It is worth noting that small players who are not intransigent can win on occasion, as the following example illustrates. If *A* is the large player and *B* and *C* are the small players in configuration 2, *B* wins when it ranks *i* fourth:

A:	*a* | *bic*	*ab* | *ic*
B:	*b* | *cai*	*bc* | *ai*
C:	*c* | *bia*	*cb* | *ia*
State:	*i*	*b*

At the second stage, *b* is supported by all three players (4 votes), whereas *a* has support only from the large player (2 votes) and *c* has support only from the two small players (2 votes).

Thus, even the small, least intransigent player can sometimes win – provided its preferences are sufficiently coincidental with those of other players – but this is the exception rather than the rule. Greater size usually helps, as I showed earlier with PR_j, and so does intransigence, especially for large players.

Yet even in configuration 1, in which size is irrelevant because there are no large or small players, ranking *i* second instead of third increases a player's wins by 160 percent (from 10 to 26). By contrast, raising *i* from fourth to third place increases a player's wins by only 25 percent (from 8 to 10), underscoring the importance of being maximally intransigent.

To get an overall picture of the effects of intransigence, sum the (i_2, i_3, i_4) values across all configurations, which gives (255, 106, 98), or proportions (0.56, 0.23, 0.21). Ranking *i* third or fourth makes little difference. But being maximally intransigent by putting *i* in second place gives a player 141 percent more wins than putting *i* in third place.

Summing wins across the different configurations, however, ignores the fact that impasse is more likely in some configurations than others, as I showed with the DR_j index earlier. But as I showed with the PR_j index, intransigence offers substantial benefits to players, large or small, in all four configurations.

A cleaner comparison – involving only large and small players in configurations 2 and 3 – is between (1) the total wins a large versus a small player obtains and (2) the total wins a player who ranks *i* second versus third obtains in these two configurations. The large–versus–small distinction gives a win ratio of 148/75 = 1.97 in favor of the larger player, which is approximately the ratio of the player weights, as noted earlier. The rank distinction (i_2 versus i_3) gives a ratio of 129/50 = 2.58 in favor of the more intransigent player, which shows that being intransigent may be even more advantageous.

Although there is no obvious standard of comparison between size and intransigence (is being twice as large comparable to ranking *i* second rather than third?), intransigence, unquestionably, is a powerful weapon in negotiations. On the other hand, the analysis assumes that all rankings below a player's most preferred alternative are equally likely, which may well be violated in practice. For example, intransigence on the part of one player may be associated with intransigence on the part of others, which may only succeed in encouraging impasse, not implementing preferred alternatives.[13] I shall consider this and other issues in the concluding section, but first I turn to a real-life application of the model.

7.6 Negotiations on world trade

Founded in Geneva in October 1947 by 23 countries, the General Agreement on Tariffs and Trade (GATT) is now an organization with 108 countries that has sponsored eight rounds of negotiations.[14] These multilateral talks have dramatically influenced international economic relations since World War II, primarily through liberalizing international trade by lowering tariff barriers.

The latest negotiations have been dubbed the "Uruguay Round" because they began in Punta del Este in September 1986. Scheduled to last four years, they collapsed in Brussels in December 1990.

As of this writing (April 1993), the full reinstatement and successful conclusion of the Uruguay Round faces formidable obstacles. The increase in bilateral trade negotiations, the rise of regional trading blocs, and a new debate about free versus fair trade all may undermine an already shaky international trading regime riddled with protectionist proclivities. But the most immediate obstacle is the conflicting interests of the three major players in the current trading round – the United States (US), the European Community (EC), and Japan (JA) – which Oxley (1990, p. 88) calls the "Big Three."

Of course, there are other influential players, most notably the so-called

[13] Socially preferred alternatives may not always be implemented either, as I illustrated with the first example in section 7.4 in which the Condorcet alternative was not selected. Of the 216 orderings for each configuration, a Condorcet alternative loses in 36 orderings of configuration 1, 17 of configuration 2, 70 of configuration 3, and 30 of configuration 4.

[14] This section is adapted in part with permission from Brams, Doherty, and Weidner (1994), which includes additional empirical material.

Cairns Group (Higgott and Cooper, 1989), which is an association of fourteen developed and developing agriculture-exporting countries that has pushed for lower agricultural barriers. Newly industrializing countries (NICs), like South Korea and Taiwan, and developing countries like Brazil, India, and the ASEAN states, also have played important roles in the Uruguay Round.

The current impasse, however, is due primarily to differences among the Big Three on agriculture, which Winham and Kizer (1990, p. 43) characterize as "the pivotal issue" and Sjöstedt (1990, p. 3) calls "the most important" unsolved problem. While less consequential to the outcome of the Uruguay Round, the issue of market access also pervades discussions of trade liberalization. Hence, I include the positions of the Big Three on this issue as well as the agricultural issue:

1 Support of agriculture through price supports or export subsidies (favored by EC, opposed by US and JA)
2 Barriers to foreign-market entry (favored by JA, opposed by US and EC).

The "barriers" in (2) do not necessarily apply to intraregional trade, such as between the United States and neighbors like Canada and Mexico. In fact, (2) more and more may be interpreted as the issue of supporting regional pacts by limiting outside access to internal markets (as EC has done), which may undermine the universality of GATT and which I shall say more about later. Issue (2), of course, is related to issue (1) if agriculture is the sector being restricted.

JA, at least for its national market, is more restrictive than either US or EC. It therefore seems fair to say that JA "favors" barriers, whereas US and, to a less extent, EC, oppose them. That the nature of JA restrictions is sometimes heavily governed by culture and practice – not comparative advantage in resources or even wages – is illustrated by the case of automobiles:

> Even if U.S. and Japanese automakers attained the same quality and production costs, U.S. producers would likely lose out. Why? Because it takes dealers to sell cars. Establishing a national dealer network from scratch in a country the size of the United States is an expensive and time-consuming task – as it is in Japan because of stratospheric real estate prices. But Japanese automakers selling in the United States don't have to build from scratch. They can piggyback onto existing GM, Ford, and Chrysler dealers because U.S. antitrust laws stipulate that producers must allow dealers to carry other lines. In contrast, by custom and because the Japanese do not enforce antitrust laws, outside firms find it extremely difficult to hook up with dealers in Japan. (Prestowitz, 1991, p. 24)

In the face of such barriers, US and EC have become increasingly less content to be unilateral free traders. For example, US and EC have used import quotas and anti-dumping provisions – sometimes in retaliation against restrictions of JA and other countries (or each other) – or they have negotiated export restraints that evade GATT rules.

At the same time, JA has not been totally recalcitrant and seems to be improving (Sanger, 1991). It has, under pressure, lifted restrictions in certain areas, like its beef market (Oxley, 1990 p. 68) and semiconductor-chip trade (Prestowitz, 1991, p. 26) with the United States; the latter agreement from 1986 to 1991 was recently extended for three years (Bradsher, 1991). Currently, however, not only does Japan have a blanket prohibition on rice imports (Farnsworth, 1990), but its rice farmers also benefit from subsidies and have, consequently, become "the most protected farmers in the world" (Passel, 1990).

Call the positions on issues (1) and (2)

A (for agricultural supports) and $\bar{A}$ (against supports)
B (for barriers) and $\bar{B}$ (against barriers).

Positions on both these issues define four possible *platforms*: AB, $A\bar{B}$, $\bar{A}B$, and $\bar{A}\bar{B}$. I assume that the players can order these platforms from best to worst, based on primary and secondary goals.

Recall from section 4.4 that a player's (1) primary goal distinguishes its two best from its two worst platforms, whereas a (2) secondary goal distinguishes between its two best platforms and between its two worst platforms. Thus, if (1) were $\bar{A}$ and (2) were $\bar{B}$, the player would order the platforms, from best to worst, as follows: ($\bar{A}\bar{B}$, $\bar{A}B$, $A\bar{B}$, AB).[15]

I assume this ordering to be the preferences of US. I summarize below the primary and secondary goals, and the preferences they imply, of the other players as well:

US: (1) $\bar{A}$ and (2) $\Rightarrow \bar{B}$ ($\bar{A}\bar{B}$, $\bar{A}B$, $A\bar{B}$, AB)
EC: (1) A and (2) $\Rightarrow \bar{B}$ ($A\bar{B}$, AB, $\bar{A}\bar{B}$, $\bar{A}B$)
JA: (1) $\bar{A}$ and (2) $\Rightarrow B$ ($\bar{A}B$, $\bar{A}\bar{B}$, AB, $A\bar{B}$).

Because of its pivotalness in the Uruguay Round, I make the issue of agricultural supports primary for all players.

It is certainly possible that the players' positions on the market-entry issue – and regional versus worldwide pacts, like GATT, which this issue raises – will assume greater importance in the future. Indeed, the ultimate failure of the Uruguay Round may lead to trade agreements by continental blocs, such as the Americas, Europe, and Asia (Passel, 1991), which some analysts view with alarm (Silk, 1991), others consider salutary (Prestowitz, 1991), but which may actually be strategic: "By preparing the ground for a series of bilateral trade deals with every country in Latin

15 See Hillinger (1971) and Kadane (1972) for an analysis of the effects of combining different positions into platforms. In the interest of clarity, I include commas between the platforms and put the ordered platforms in parentheses.

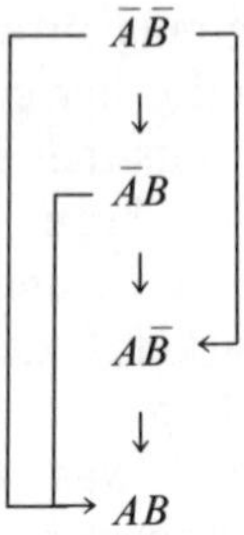

Note: Arrows emanating from a higher platform to a lower platform indicate that the former is *socially* preferred (i.e., by a majority of two out of three players) to the latter.

Figure 7.3 Social hierarchy of majority preferences for platforms

America," the United States and its potential partners may be "quietly hedging their bets" ("Hedging," 1990). Conceivably, however, the "mini-lateralism" of such blocs may evolve into the multilateralism of GATT, facilitating rather than undermining world trade (Yarbrough and Yarbrough, 1992), though this point is hotly contested (Uchitelle, 1991).

Observe that each of the three players has a different first, second, third, and last preference, suggesting a lack of social consensus. Nevertheless, by comparing, for each pair of platforms, which is *socially* preferred (i.e., by a majority of two of the three players), one obtains the social preference ordering shown in figure 7.3. (Later I shall indicate that social preferences based on majority rule may not be descriptive of how negotiation outcomes on world trade are determined.)

Notice that two of the three players, indicated by the three arrows emanating from $\bar{A}\bar{B}$, prefer $\bar{A}\bar{B}$ to each of the other three platforms. The fact that these other platforms can also be ordered so that all social preference relations flow "downward" from $\bar{A}\bar{B}$ to $\bar{A}B$ to $A\bar{B}$ to AB establishes the existence of a *social hierarchy* of platforms (so there is no paradox of voting, or cyclical majorities), with $\bar{A}\bar{B}$ the Condorcet alternative because it defeats the other three alternatives in separate pairwise contests. In this case, social preferences are said to be *transitive*.

Even $\bar{A}\bar{B}$, however, if pitted against each of the other platforms, would always be opposed by one of the three players. Thus, if each of the players is able to veto the social choice of a platform, the fact that there exists a social hierarchy based on majority preferences does not establish that a social consensus will develop around $\bar{A}\bar{B}$ at the top of the hierarchy. Quite the contrary: each of the platforms would be vetoed by the player who prefers another. Hence, if unanimous consent is required (as is probably the case among the Big Three in the Uruguay Round), it will not be achieved.

Before applying the negotiation model to the preferences of each player, some assumptions must be made about where impasse, which I capitalize in this section as "*I*," falls in the orderings of the players. Assume EC is adamant in its position on (1); unalterably opposed to $\bar{A}$, it insists on agricultural supports and so puts *I* in third place.

Assume US puts *I* in fourth place: it will not give up on both its primary and secondary goals. Then the only way that unanimous consent can be achieved is if JA puts *I* in fifth place. If this is the case, support of the various alternatives will evolve to the following (I do not show the earlier stages):[16]

US: $(\bar{A}\bar{B}, \bar{A}B, A\bar{B}, I \mid AB)$
EC: $(A\bar{B}, AB, I \mid \bar{A}\bar{B}, \bar{A}B)$
JA: $(\bar{A}B, \bar{A}\bar{B}, AB, A\bar{B} \mid I)$.

By the time the players have lowered their support to the points indicated by the vertical bars, there will be unanimous support for $A\bar{B}$.

Yet both $\bar{A}\bar{B}$ and $\bar{A}B$ beat $A\bar{B}$ in the hierarchy! The former platforms "lose," once *I* is inserted in the preference rankings, because of EC's hypothesized intransigence – it will not lower its support below *I* because, by so doing, it cannot effect a better outcome at any stage. The somewhat diminished intransigence hypothesized for US – and still less for JA – ensures that $A\bar{B}$ rather than $\bar{A}\bar{B}$ will be the outcome (the reverse would be the case if JA ranked *I* fourth and US ranked it fifth). In this manner, a player's higher placement of *I* induces the choice of a preferred platform for itself, even though this platform may fall lower in the social hierarchy, based on majority-rule voting, than others, and hence not be the Condorcet alternative.

Time will tell, if this hypothetical attribution of platform preferences is correct and the unanimity rule is operative, whether EC ranks *I* higher than US and JA. The latter players' rankings of *I* matter, too, as I have just shown, and also may not be as hypothesized. Indeed, the fact that the Big Three have disagreed since 1990 suggests that one or more of these players ranks *I* higher than hypothesized.

The hypothesized preferences of the Big Three (for *I* as well as for the different platforms), the unanimity decision rule, and negotiations that unfold in the manner of the model certainly do not capture all the nuances

16 Because the Big Three surely know each others' positions very well on the two major issues I've discussed, the incomplete-information assumption may seem quite unreal. On the other hand, their disagreement is ultimately over specifics, and where fallback positions lie on these, which is private information about which only good guesses can usually be made. While the model might be more appropriately applied to these specifics, I am not privy to them.

of any endgame that may be played out in the Uruguay Round and later. Other preferences and rules, and even new players, might be incorporated into the analysis if it is believed that they offer a more realistic portrayal of the current trade-negotiation game.

It is the incomplete-information framework, based on the revelation assumption (RA), that drives the analysis. It offers an enlightening way of viewing the sequential revelation of positions, and possible changes in support patterns, as players offer compromises – specifically, by progressively supporting lower-ranked platforms when, according to rules 1–4, it is rational to do so.[17] A model in which players respond to each other over time is needed, I believe, to explicate the dynamics of multilateral negotiations in larger games.

7.7 Intransigence in negotiations

Intransigence is a powerful but dangerous card for a negotiator to play in multilateral negotiations. If other parties are also likely to hold out by not retreating to fallback positions, then impasse can be expected.

To be sure, there is nothing irrational about this outcome if players prefer it to other alternatives. Furthermore, it may be effective in inducing other parties to be more forthcoming. On the other hand, if one player's intransigence fosters intransigence on the part of others, such recalcitrant behavior can lead to a disaster for all.

Although one player's intransigence does not depend on another's in the negotiation model, in principle it can be made endogenous. The direction of the dependence might vary from situation to situation. Thus, a player might conceivably react to others' intransigence either by turning recalcitrant itself or becoming more accommodating.

Presumably, a player's reputation from earlier play will affect when others view its intransigence as credible. This credibility may also be enhanced by immediate actions, such as "burning bridges" to demonstrate its resolve. A distinguishing feature of the negotiation model, besides its dynamic character, is its inclusion of a preference for impasse as well as for the regular alternatives. Cooperative game-theoretic models of bargaining often include a "disagreement point," but this is less common in noncooperative models.

The main theoretical finding is that the more acceptable that impasse is

[17] I am hampered, however, in having insufficient information to test whether support patterns on world trade issues evolved in the manner predicted by the model. In fact, I am not even sure what the "true" preferences of the players are, which is why I claim only to have illustrated a framework, not rigorously tested a model. On the other hand, when preference information is sparse, one can do sensitivity analysis, testing how dependent one's conclusions are on different plausible assumptions about the preferences.

to a player, the more it will achieve its preferred outcome on the average, usually by a factor of two or more when it raises *i* from third to second place. Overall, intransigence may be a more potent force than size – measured by a player's ability to affect the outcome of negotiations – but the two factors are not directly comparable.

Although intransigence works best in combination with size, small players can sometimes achieve their best outcome if their preferences coincide sufficiently with those of others. Thus, the actual game being played is crucial in determining whether intransigence is helpful or not.

Multilateral negotiations often can be reduced to bilateral negotiations, or a series of bilateral negotiations, but sometimes they resist such a simplification. This seems to be the case in world trade negotiations, in which there are at least three major players, making a reduction of the conflict to a two-person game difficult to defend.

The positions of the Big Three on agriculture, the most contentious issue in the Uruguay Round, have been divergent. But the issue of market access also is salient, even if it is less relevant to the Uruguay Round but instead part of a larger game.[18] The latter issue, should it become paramount, may be instrumental in restructuring the players into regional blocs and altering the fundamental nature of the international trading game.

The analysis of the game among the Big Three, reflecting their positions on both agriculture and market access, indicates that not only do their preferences differ but also demonstrates that one or more will have to make significant compromises to reach a consensus. If not, failure will result because an agreement, at least in the eyes of some players, will be worse than no agreement. This analysis, in my opinion, illuminates why this is the case and thereby clarifies the rational foundations of disagreement in negotiations.

As I indicated in sections 7.1 and 7.2, the negotiation model in this chapter departs significantly from the theory I developed in earlier chapters. But in larger games, simplifying assumptions seem necessary to make the analysis tractable, and a model that players might actually use realistic.

In the case of the negotiation model, I have restricted the information available to the players about each other's preferences, which they discover only in stages. Using the revelation assumption (RA), they act on this information by retreating to fallback positions or choosing impasse. Their initial lack of information, and their discovery of new information, seem plausible assumptions to make in many negotiations.

That intransigence "helps" may come as no surprise, but it is revealing how much it compensates for size. Indeed, it may even be a substitute for

[18] Clearly, the identity of the players, and what games are being played, is sometimes a mystery that may take empirical investigation to unravel.

greater size, enabling smaller players, by being inflexible, to compete with larger more flexible players.

Perhaps more surprising than the effects of size and intransigence is that the negotiation process may lead to the selection of a non-Condorcet alternative. This result seems especially unfortunate when impasse is the outcome, yet a majority of players prefer a regular alternative not only to impasse but also to the other regular alternatives. Indeed, this situation may describe world trade negotiations today.

8 Summary and conclusions

I begin by summarizing the main themes in each chapter and then offer some concluding remarks on TOM and its applications:

Chapter 1

I distinguished a game configuration, which gives the basic structure of payoffs in a matrix, from a game, which also specifies the initial state. Using game configuration 56 as an example (I usually dropped "configuration"), I illustrated the standard concepts of game theory – Nash equilibrium, dominant strategy, complete information, and so on. I then gave the rules of play and rationality rules of TOM, emphasizing that payoffs, which are strictly ordinal, accrue only when a final state, or outcome, is reached in the move–countermove process.

I illustrated the consequences of backward induction from each state in game 56, assuming a game never returns to its initial state and that the players make two-sided rationality calculations, always taking into account the other player's rational choices. Game 56 was shown to have three nonmyopic equilibria (NMEs), making the outcome highly sensitive to the initial state.

Applied to the Samson and Delilah story, beginning when Samson is unforthcoming about the source of his strength and Delilah has not yet nagged him, TOM correctly predicts that Delilah will start nagging Samson and he will succumb to her entreaties. I indicated that certain moves in games, such as Samson's reversing himself after revealing the secret of his strength, may be infeasible, which must be taken into account in applying TOM.

Chapter 2

Introducing additional game-theoretic concepts into the analysis, I argued that most real-life games are not games of total conflict and have, as a consequence, Pareto-inferior states. Elections, however, are important exceptions. I suggested, and illustrated with examples from recent presidential elections, how the three ordinal games of total conflict – two of which have one NME and the third two NMEs – can be used to model different kinds of races between an incumbent and a challenger, or two challengers.

I then defined and illustrated anticipation games (AGs), which have as their entries the NMEs into which each states goes. If there is more than one NME, one can readily determine which, if any, are Nash equilibria in the AGs.

In several games, which I illustrated by the biblical story describing the pursuit of the Israelites by Pharaoh, there are no pure-strategy Nash equilibria in their AGs, which creates incentives to move in an unstable environment. Indeed, Pharaoh, after deciding to let the Israelites flee Egypt, retracted on his promise. In the game used to model this biblical story, the precise moves the players made, which culminate in the NME favoring the Israelites, are predicted by TOM.

I concluded by describing the anticipation problem in literature, epitomized by Faulkner's *Light in August*, in which mixed strategies, which may be optimal in two-person zero-sum games in the standard theory, seem to have been invoked. Whereas the ordinal framework of TOM does not admit mixed strategies, it does highlight those AGs in which there are two NMEs and in which neither player has a dominant strategy. This is precisely the kind of AG that gives rise to an anticipation problem.

Chapter 3

An apparent conflict between rules 5 and 6 of TOM was resolved by the two-sidedness convention (TSC), which prescribes that a player, who can do better by not moving from a state, nevertheless will move if the other player's (rational) move induces a state Pareto-inferior to that which the first player can induce by moving. Consequently, it may be in both players' interest that one player make a "sacrifice," lest it do still worse by not moving.

I illustrated the set of games to which TSC is applicable with a mugging game, in which it is rational for the victim voluntarily to submit before the mugger resorts to force. Especially when threatened with a gun, statistics on robberies indicate that victims tend to be rational (i.e., magnanimous) in this regard.

I then developed a generic Magnanimity Game (MG), which subsumes 12 specific ordinal games, and showed that, depending on the game, each of the four states may be an NME. (Thus, differing views in the international relations literature about the rationality of magnanimity are not necessarily in conflict – what is rational depends on the specific MG game being played.) I summarized rational behavior in MG with three propositions and then illustrated the different possible outcomes by describing the aftermaths of several nineteenth- and twentieth-century wars.

Chapter 4

I dropped the prohibition against cycling and, based on new rationality rules (5′ and 6′), showed that if a game is cyclic because no player ever reaches a state best for itself when it has the next move, one player may induce a preferred outcome if it has moving power. (Moving power is the ability to continue moving when the other player must eventually stop; it is defined only for cyclic games, which I showed can cycle either in a clockwise or a counterclockwise direction, but not in both directions.) Moving power, however, is not always helpful: there are games in which a player with moving power will prefer to stop first itself. In such games, moving power is irrelevant, because it does not make a difference which player possesses it. In a few games it may actually be counterproductive to exercise moving power – it would be better for the other player to do so – and therefore it is ineffective.

On the other hand, when the possession of moving power does make a difference in implementing preferred outcomes, it is said to be effective. Egypt's moving power in its running conflict with Israel from 1948 to 1979 appears to have been effective, inducing Israel after five wars, all of which it won militarily, to join in a compromise settlement.

I also used moving power to illustrate the advantage it confers on a superior being (*SB*) in the Revelation Game, in which *SB* is able to induce belief on the part of a person (*P*) but would prefer not to do so. I suggested that this conflict leads to an unstable situation in which periods of belief, instilled by revelation, alternate with periods of doubt sowed by nonrevelation.

Finally, I modeled the use of bombing by the United States in the Vietnam war as an attempt to exert its moving power, necessitated at the end of the war by an apparent misperception of U.S. preferences by North Vietnam. Although the Linebacker II campaign was probably instrumental in the signing of a peace treaty in 1973, its success was ephemeral. This treaty was abrogated in 1975, after U.S. forces were withdrawn from South Vietnam and North Vietnam successfully invaded the South, illustrating how moving power may slip away.

Chapter 5

Two new kinds of power, order power and threat power, were defined and illustrated. Order power is applicable to games with indeterminate states, in which the NME that is induced from that state depends on which player moves from it first.

From some indeterminate states a player would prefer to move first, whereas from others it would prefer that the other player move first. In either event, a player with order power can dictate the order of moves, so it always benefits, rendering this power effective. I compared the effects of order power in game 56 with those of moving power and showed that, in the Samson and Delilah story, order power would enable Delilah to attain her preferred NME from the mutually worst state in this game.

Threat power comes in two varieties, compellent and deterrent, and starts with the choice, or threatened choice, of strategies, not with moves from a state. It requires prior communication (PC) and is applicable only to repeated or continuing games, in which reputation matters and a player may therefore be willing to suffer a Pareto-inferior outcome – by carrying out a threat – to make its future threats credible.

I showed how deterrent threats reinforce the cooperative state in Prisoners' Dilemma and undermine it in Chicken. Following a theoretical analysis of threat power, I gave results on compellent and deterrent threats for other games as well.

Chicken and one other game were used as models of the Cuban missile crisis, in which threats figured prominently. I also analyzed the use of threats in the Polish crisis of 1980–1, first by Solidarity and then by the Communist party. I concluded with a discussion of the varieties of power that players may try to exploit, arguing that each of the three kinds analyzed may come into play in different situations.

Chapter 6

I explored different consequences of incomplete information, which players may have either about the preferences of an opponent or about its power (relative to theirs). The same game configuration used to model the Polish crisis seemed also to capture the conflict between the Union and Confederacy prior to the outbreak of the American Civil War in 1861.

The preferences of both sides for the different states seem not to have been in dispute, but who might prevail in a war was less clear. Although some Southern leaders recognized their military inferiority, they hoped that, by striking the first blow, they could demonstrate convincingly to the North their commitment to slavery and thereby their threat power. But they seemed to have underestimated Lincoln's resolve; it was the Union's threat outcome, after a long and costly war, that was implemented.

Who possessed threat power was not the issue in the Iran hostage crisis of 1979–81. Rather, it was President Carter's misperception of Ayatollah Khomeini's preferences, which led him to believe, erroneously, that he could implement his preferred NME in this game with a deterrent threat. By contrast, the outcome that incomplete information may have facilitated in the Cuban missile crisis was salutary for both President Kennedy and Premier Khrushchev, but it is not clear whether it was induced by deception on the part of Khrushchev or simply by a change in his preferences brought on by the frightful consequences of nuclear war.

Signaling in an environment of incomplete information is central to the plot of Shakespeare's *Hamlet*, which ends in the deaths of the two principals, Hamlet and Claudius. Because the game I used to model their conflict has a unique NME, it is hard to see how these figures could have avoided their tragedy. In fact, it seems that Hamlet was less a victim of indecisiveness – often pointed to as his tragic flaw – than of the very nature of the game in which he was caught up. True, he had a hand in defining it, but once involved, his actions to accumulate incriminating evidence against Claudius, and Claudius' attempts to discover whether Hamlet was plotting against him, were fiendishly clever. Their tragedy cannot be pinned on their ineptitude but is, rather, attributable to the exigencies of the strategic situation that embroiled them.

I supplemented the foregoing examples with an analysis of MG in the case of incomplete information. More specifically, I showed exactly what information each player would need to acquire about its opponent's preferences to act rationally in this generic game.

Chapter 7

The three players in the negotiation model developed in this chapter were assumed to possess no information about the preferences of the other two, except as it is revealed in the course of play when the players progressively invoke fallback positions to try to reach a consensus (assumed to be a simple majority of their "votes," or weight). Thus, the players are not able to make nonmyopic calculations, which is a significant departure from the earlier models that I made for reasons of both tractability and realism – often players in larger games *are* in the dark about where the other players stand in negotiations. Nevertheless, players do respond to each other's choices, which mirror the dynamics of TOM.

I illustrated the dynamics of negotiation initially with an example that included a player with a tie-breaking vote (i.e., the chair). By retreating to their fallback positions, the players eventually select the chair's best alternative. I then analyzed three-person negotiations in which I postulated a preference of the players for impasse, as well as for three regular alternatives, with the negotiations governed by some simple rules of play.

The main result was that intransigence, or an unwillingness to resort to fallback positions because impasse is preferred, generally works to a player's advantage. Greater size, which translates into an enhanced ability to effect a preferred outcome, also helps, but intransigence is a potent force by itself.

I illustrated the analysis by applying it to negotiations among the United States, the European Community, and Japan on two issues related to free trade. Somewhat surprisingly, a Condorcet alternative preferred by two of the three players may not be selected when negotiations proceed according to the model. In the currently suspended Uruguay Round of GATT, the outcome has so far been impasse, which only one of the three players – or even none – may prefer to some agreement.

Concluding remarks

One of the features of TOM that I singled out in the Introduction is the distinctive point of view that it provides. Among other things, the theory is ordinal: it assumes no quantitative calculations, based on cardinal utilities and probabilities that can be used to define expected values, which often are intrinsically unmeasurable. Also, play starts in a state, not with strategy choices or announcements (except when threats or deception are used), and outcomes are contingent on these.

To be sure, utilities and quantitative calculations have their place, but not usually as a first approximation in analyzing questions of strategy. Rather, there seem to me more fundamental issues, tied to the following three concepts:

*1 Stability. A*n NME is the basic equilibrium concept used, under the assumption that most real-life players are not so myopic – especially when they make important decisions – as to consider only the immediate effects of a departure from a state, without taking into account possible responses of other players as well as themselves. NMEs always exist; when there is more than one, the selection of one depends on the initial state and sometimes on order power, which helps to solve the multiple-equilibrium problem of standard game theory. (This problem is compounded by the existence of Nash equilibria in mixed strategies, which TOM precludes because of its ordinal character.)

At the same time, players may contemplate the possibility of cycling back to the initial state, which NMEs prohibit, in an effort to outlast an opponent. In fact, I relaxed the prohibition against cycling in cyclic games, in which termination may depend on which player, if either, possesses moving power.

2 Power. Many if not most games are between players with different capabilities, so it is appropriate to consider different kinds of asymmetries that may occur in their play. Moving, order, and threat power reflect, respectively, the ability to hold out longer in a continuing conflict, to control who moves first from an indeterminate state, and to choose a strategy that will compel or deter untoward future choices when a game is repeated. These different kinds of power may reinforce or undermine the stability of an NME. Intransigence, or a willingness to accept impasse, may also be considered a kind of power in the negotiation model developed in chapter 7.

3 Information. Players may not have complete information about either an opponent's preferences or its power, so it is important to analyze the effects of incomplete information on the play of a game. I gave examples of how a lack of information led to misperceptions, possible deception, and even war, though occasionally playing the "wrong" game may actually be helpful. Surprisingly, omniscience may be harmful if an opponent is aware that a player has such predictive abilities.

The more general point is that information at varying levels can be formally incorporated into the TOM analysis. Prescriptively speaking, this analysis may help players determine whether they can make rational strategy choices without acquiring additional information or should, instead, search for such information.

Looking through the lenses of these three concepts, TOM addresses central issues in a parsimonious fashion. Grounded in only a few rules of play, its theoretical foundations are simple but generate a host of provocative consequences.

Although doing a detailed analysis of even a few 2×2 games can be taxing, I recommend this exercise to gain an appreciation of the calculations underlying the concepts. This done, however, there is no reason not to turn to the Appendix to look up results for any of the 57 2×2 conflict games.

I illustrated the application of TOM to two larger games in the Introduction, which gave answers different from the standard theory. But I am ambivalent about extending the analysis in rote fashion to more complex games. While backward induction can be carried out quickly on a computer, it would take some effort to develop an efficient algorithm for making nonmyopic calculations in larger matrix games. But do people really think through the manifold choices in larger games and impeccably choreograph their moves in them? Or, plagued by incomplete information, are they more likely to follow simple rules of thumb, like those assumed in the spare negotiation model of chapter 7?

I presented the negotiation model in part to provide an alternative perspective on how players might make their moves in more complex, inchoate environments. Surely other plausible rules of play can be postulated that shed light on different situations.

There is no science for formulating "best" rules. This problem falls in the realm of modeling, which is more an art than a science. It relies on good intuition and a familiarity with the strategic choices players make in the empirical situations one is trying to explain.

I have deliberately gone back and forth between the theory and applications in trying to develop TOM. Clearly, the theory can be applied in new empirical domains, which should help to motivate further refinements. I strongly urge a continuing dialogue between theory and applications to prevent TOM from becoming sterile, baroque, or irrelevant.

Appendix

There are 78 structurally distinct 2×2 strict ordinal games in which the two players, each with two strategies, can strictly rank the four states from best to worst. These games are "distinct" in the sense that no interchange of the column player's strategies, the row player's strategies, the players, or any combination of these can transform one game into any other. That is, these games are structurally different with respect to these transformations.

Of the 78 games, 21 are no-conflict games with a mutually best (4,4) state. These states are always Nash and nonmyopic equilibria (NMEs) in these games; no kind of power – moving, order, or threat – is needed by either player to implement them as outcomes.

I list here the remaining 57 games, in which the players disagree on a most-preferred state. The numbers used in the original listing of the 78 games by Rapoport and Guyer (1966) are given in parentheses after the numbers used here.[1] The 57 games are divided into three main categories: (i) those with one NME (31 games), (ii) those with two NMEs (24 games), and (iii) those with three NMEs (2 games). I have grouped together at the end of the list the nine games with indeterminate states – seven of which fall in category (ii) and two of which fall in category (iii) – in which order power is effective when play starts at an indeterminate state.

Moving power outcomes that the row and column players can induce are indicated by superscripts in the key given below. If the outcomes induced by moving power are different, then this power is effective if the player who possesses it can implement a better outcome for itself than if the other player possessed it; otherwise, it is irrelevant or ineffective, which

[1] Another complete listing of the 78 games is given in Brams (1977), in which the games are divided into three categories based on their vulnerability to deception. The moving-power outcomes identified in the 57 conflict games listed in Brams (1982a, 1983) differ somewhat from those given here because of changes I have made in the rules of play of TOM. Threat-power outcomes were previously identified in Brams (1983, 1990) and Brams and Hessel (1984). Order power is a new concept, which I use instead of staying power (Brams, 1983; Brams and Hessel, 1983; Kilgour, De, and Hipel, 1987) for reasons given in section 5.1

are categories distinguished in tables 4.3, 4.4, and 4.5. (These three tables give, respectively, cyclic games that are strongly, moderately, and weakly cyclic.) All games not in these tables, which are the games in the listing below in which moving power outcomes are not identified, are noncyclic.

Threat power outcomes are also indicated by superscripts in the key given below, with compellent threat outcomes distinguished from deterrent threat outcomes. Threat power is always effective when the row and column player can induce different outcomes; when they induce the same outcome, threat power is irrelevant.

Order power is defined only in games with indeterminate states, which are identified in their anticipation games (AG) given in the listing below. Order power always leads to different outcomes and is invariably effective.

Certain games are identified by their descriptive names or by the conflicts modeled by them in the text. These include Samson and Delilah (56), Total conflict (11, 25, and 44), Pursuit of Israelites (42), Revelation (48), Vietnam bombing (50 and 37), Prisoners' Dilemma (32), Chicken (57), Cuban missile crisis (21, 30, and 57), Polish crisis, 1980–1 (22), Union–Confederacy conflict (22), Iran hostage crisis (5 and 50), and Hamlet–Claudius conflict (26).

The key to the symbols is as follows:

(x,y) = (payoff to Row, payoff to Column)

$[a,b]$ = [payoff to Row, payoff to Column] in anticipation game (AG)

$[w,x]/[y,z]$ = indeterminate state in AG, where $[w,x]$ is the NME induced if Row has order power, $[y,z]$ if column has order power, from corresponding state in original game

4 = best; 3 = next best; 2 = next worst; 1 = worst

Nash equilibria in original game and AG underscored (except when there is only one NME in AG)

NMEs in original game circled

m/M = moving-power outcome row/column can induce

t/T = threat-power outcome row/column can induce

c/d = compellent/deterrent threat outcome

(i) 31 games with one NME

1 (13)

(3,4)^{mMTc}	(4,2)
(2,3)	(1,1)

2 (14)

(3,4)^{mMTc}	(4,2)
(1,3)	(2,1)

3 (15)

(3,4)^{mMTc}	(4,1)
(2,3)	(1,2)

Cuban missile crisis
4 (16)

(3,4)^{mMTc}	(4,1)
(1,3)	(2,2)

Iran hostage crisis
5 (17)

(2,4)^{Tc}	(4,2)
(1,3)^{m}	(3,1)^{M}

6 (18)

(2,4)^{Tc}	(4,1)
(1,3)^{m}	(3,2)^{M}

7 (7)

(3,3)	(4,2)
(2,4)	(1,1)

8 (8)

(3,3)	(4,2)
(1,4)	(2,1)

9 (9)

(3,3)	(4,1)
(1,4)	(2,2)

10 (10)

(2,3)	(4,2)
(1,4)	(3,1)

Total conflict
11 (11)

(2,3)	(4,1)
(1,4)	(3,2)

12 (40)

(3,4)^{mMTc}	(4,1)
(2,2)	(1,3)

13 (41)

(3,4)^{mMTc}	(4,1)
(1,2)	(2,3)

14 (31)

(3,4)^{Tc}	(2,2)
(1,3)	(4,1)

15 (32)

(3,4)^{Tc}	(2,1)
(1,3)	(4,2)

16 (33)

(3,4)^{Tc}	(1,2)
(2,3)	(4,1)

17 (34)

(3,4)^{Tc}	(1,1)
(2,3)	(4,2)

18 (35)

(2,4)^{Tc}	(3,2)
(1,3)	(4,1)

19 (36)

(2,4)^{Tc}	(3,1)
(1,3)	(4,2)

20 (37)

(3,4)^{Tc}	(2,3)
(1,2)	(4,1)

Cuban missile crisis
21 (38)

(3,4)^{Tc}	(1,3)
(2,2)	(4,1)

Polish crisis, 1980–1
Union–Confederacy crisis
22 (39)

(2,4)^{Tc}	(3,3)^{td}
(1,2)	(4,1)

23 (42)

(3,3)^{mMTc}	(4,1)
(2,2)	(1,4)

24 (43)

(3,3)^{mMTc}	(4,1)
(1,2)	(2,4)

Total conflict
25 (45)

(3,2)^{M}	(4,1)
(2,3)^{m}	(1,4)

Hamlet–Claudius conflict
26 (46)

(3,2)^{M}	(4,1)
(1,3)^{m}	(2,4)

27 (47)

(2,3)	(4,1)
(1,2)	(3,4)^{mMTd}

28 (48)

(2,2)	(4,1)
(1,3)	(3,4)^{mMTd}

Cuban missile crisis

29 (72)

(3,2)	(2,1)
$(4,3)^{mMtdTc}$	(1,4)

30 (77)

(2,2)	(4,1)
$(3,3)^{mMtdTc}$	(1,4)

31 (78)

(2,2)	(3,1)
$(4,3)^{mMtdTc}$	(1,4)

(ii) 24 games with two NMEs

Prisoners' Dilemma

32 (12)

(2,2) [2,2]	(4,1) [3,3]
(1,4) [3,3]	$(3,3)^{tdTd}$ [3,3]

33 (19)

$(3,4)^{MTc}$ [3,4]	$(4,3)^{mtd}$ [4,3]
(1,2) [3,4]	(2,1) [3,4]

34 (20)

$(3,4)^{MTc}$ [3,4]	$(4,3)^{mtd}$ [4,3]
(2,2) [3,4]	(1,1) [3,4]

35 (21)

$(2,4)^{Tc}$ [2,4]	$(4,3)^{mtd}$ [4,3]
(1,2) [2,4]	$(3,1)^{M}$ [2,4]

Vietnam bombing

36 (49)

$(3,4)^{MTc}$ [3,4]	$(4,3)^{mtd}$ [4,3]
(2,1) [3,4]	(1,2) [3,4]

37 (50)

$(3,4)^{MTc}$ [3,4]	$(4,3)^{mtd}$ [4,3]
(1,1) [3,4]	(2,2) [3,4]

38 (51)

$(3,4)^{MTc}$ [3,4]	$(4,2)^{m}$ [4,2]
(2,1) [3,4]	(1,3) [3,4]

39 (52)

$(3,4)^{MTc}$ [3,4]	$(4,2)^{m}$ [4,2]
(1,1) [3,4]	(2,3) [3,4]

40 (53)

$(3,3)^{MTc}$ [3,3]	$(4,2)^{m}$ [4,2]
(2,1) [3,3]	(1,4) [3,3]

41 (54)

$(3,3)^{MTc}$ [3,3]	$(4,2)^{m}$ [4,2]
(1,1) [3,3]	(2,4) [3,3]

Pursuit of Israelites

42 (73)

$(2,4)^{M}$ [2,4]	(4,1) [3,2]
$(3,2)^{m}$ [3,2]	(1,3) [2,4]

43 (74)

$(2,4)^{M}$ [2,4]	(3,1) [4,2]
$(4,2)^{m}$ [4,2]	(1,3) [2,4]

Total conflict

44 (75)

$(2,3)^{M}$ [2,3]	(4,1) [3,2]
$(3,2)^{m}$ [3,2]	(1,4) [2,3]

45 (76)

$(2,3)^{M}$ [2,3]	(3,1) [4,2]
$(4,2)^{m}$ [4,2]	(1,4) [2,3]

46 (70)

$(3,4)^{MtcTd}$ [3,4]	(2,1) [4,2]
$(4,2)^{m}$ [4,2]	(1,3) [3,4]

47 (71)

$(3,3)^{MtcTd}$ [3,3]	(2,1) [4,2]
$(4,2)^{m}$ [4,2]	(1,4) [3,3]

Revelation
48 (57)

(2,3) [3,4]	(4,2)^{m} [4,2]
(1,1) [3,4]	(3,4)^{MTd} [3,4]

9 games with indeterminate states

49(44)

(2,4)^{Tc} [3,3]	(4,1) [2,4]
(1,2) [2,4]/[3,3]	(3,3)^{mMtc} [2,4]

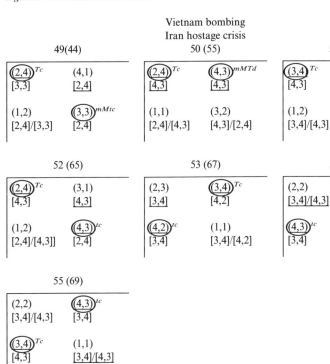

Vietnam bombing
Iran hostage crisis
50 (55)

(2,4)^{Tc} [4,3]	(4,3)^{mMTd} [4,3]
(1,1) [2,4]/[4,3]	(3,2) [4,3]/[2,4]

51 (64)

(3,4)^{Tc} [4,3]	(2,1) [3,4]/[4,3]
(1,2) [3,4]/[4,3]	(4,3)^{tc} [3,4]

52 (65)

(2,4)^{Tc} [4,3]	(3,1) [4,3]
(1,2) [2,4]/[4,3]]	(4,3)^{tc} [2,4]

53 (67)

(2,3) [3,4]	(3,4)^{Tc} [4,2]
(4,2)^{tc} [3,4]	(1,1) [3,4]/[4,2]

54 (68)

(2,2) [3,4]/[4,3]	(3,4)^{Tc} [4,3]
(4,3)^{tc} [3,4]	(1,1) [3,4]/[4,3]

55 (69)

(2,2) [3,4]/[4,3]	(4,3)^{tc} [3,4]
(3,4)^{Tc} [4,3]	(1,1) [3,4]/[4,3]

(iii) 2 games with three NMEs

Samson and Delilah
56 (56)

(2,4)^{Tc} [3,3]	(4,2)^{m} [4,2]
(1,1) [2,4]/[3,3]	(3,3)^{Mtc} [2,4]

Chicken
Cuban missile crisis
57 (66)

(3,3) [3,3]	(2,4)^{Tc} [4,2]/[3,3]
(4,2)^{tc} [3,3]/[2,4]	(1,1) [2,4]/[4,2]

Glossary

This glossary contains definitions of game-theoretic, move-theoretic, and related terms used in this book, which are described in relatively nontechnical language. More extended and rigorous definitions of some concepts can be found in the text.

Anchor. An anchor is the endstate from which backward induction proceeds, which occurs after one complete cycle in the case of a nonmyopic equilibrium.

Anticipation game (AG). An anticipation game is described by a payoff matrix, whose entries, given in brackets, are the nonmyopic equilibria into which each state of the original game goes.

Approval voting. Approval voting is a voting procedure in which voters can vote for (i.e., approve of) as many alternatives as they like in an election with more than two alternatives; the alternative with the most votes wins.

Backtracking. Backtracking occurs when a player moves first in one direction and then reverses the direction of its move.

Backward induction. Backward induction is a reasoning process in which players, working backward from the last possible move in a game, anticipate each other's rational choices.

Blockage. Blockage occurs when it is not rational, based on backward induction, for a player to move from a state.

Breakdown state/strategy. A breakdown state is the Pareto-inferior state that a threatener threatens to implement, by choosing its breakdown strategy, unless the threatenee accedes to the threat state.

Cardinal utility. See Utility.

Cheap talk. Cheap talk is costless preplay communication that is not binding on the player that makes it.

Common knowledge. Players in a game have common knowledge when they share certain information, know that they share it, know that they know that they share it, and so on *ad infinitum*.

Compellent threat. In repeated play of a two-person game, a threatener's compellent threat is a threat to stay at a particular strategy to induce the threatened player to choose its (as well as the threatener's) best state associated with that strategy.

Complete information. A game is one of complete information if each player knows the rules of play, the rationality rules, the preferences or payoffs of every player for all possible states, and which player (if either) has moving, order, or threat power. When this is not the case, information is incomplete. See also Perfect information.

Condorcet alternative. A Condorcet alternative is an alternative that defeats every other alternative, based on majority rule, in separate pairwise contests.

Configuration. See Game configuration; Weight configuration.

Conflict game. A conflict game is a 2×2 strict ordinal game in which there is no mutually best (4,4) state; a no-conflict game is one in which there is such a state.

Constant-sum (zero-sum) game. A constant-sum (zero-sum) game is a total-conflict game in which the utility payoffs to the players at every outcome sum to some constant (zero); if the game has two players, what one player gains the other player loses.

Contingency. A contingency is the set of strategy choices made by players other than the one in question.

Cooperative game. A cooperative game is a game in which the players can make binding and enforceable agreements, usually with respect to how the payoffs will be split among them.

Cyclic game. A 2×2 strict ordinal game is a cyclic if moves either in a clockwise or a counterclockwise direction never give a player its best payoff when it has the next move. A cyclic game is strongly cyclic if each player always does immediately better by moving; otherwise, it is either moderately or weakly cyclic.

Cyclical majorities. Cyclical majorities occur when majorities of voters prefer alternative x to y, y to z, and z to x, indicating the lack of a transitive social ordering; such majorities exist when there is a paradox of voting.

Deception strategy. In a two-person game of incomplete information, a deception strategy is a player's false announcement of a preference in order to induce the other player to choose a strategy favorable to the deceiver.

Decision rule. A decision rule specifies the subsets of actors (e.g., a simple majority in a voting body) that can take collective action that is binding on all the actors.

Determinacy ratio. The determinacy ratio is the ratio of the total number of wins of regular alternatives to the total number of wins of both regular alternatives and impasse in the negotiation model.

Deterrent threat. In repeated play of a two-person game, a threatener's deterrent threat is a threat to move to another strategy to induce the threatened player to choose a state, associated with the threatener's initial strategy, that is better for both players than the state threatened.

Dominant strategy. A dominant strategy is a strategy that leads to outcomes at least as good as those of any other strategy in all possible contingencies, and a better outcome in at least one contingency. A strictly dominant strategy is a dominant strategy that leads to a better outcome in every contingency.

Dominated strategy. A dominated strategy is a strategy that leads to outcomes no better than those of any other strategy in all possible contingencies, and a strictly worse outcome in at least one contingency. A strictly dominated strategy is a dominated strategy that leads to a worse outcome in every contingency.

Effective power. Power is effective when possessing it induces a better outcome for a player in a game than when an opponent possesses it; when the opposite is the case, power is ineffective.

Equilibrium. See Nash equilibrium; Nonmyopic equilibrium (NME).

Expected payoff. An expected payoff is the weighted sum of the payoff that a player receives from each outcome multiplied by the probability of its occurrence.

Extensive form. A game in extensive form is represented by a game tree in which the players make sequential choices but do not necessarily know all the prior choices of the other players.

Feasibility. A move is feasible if it can plausibly be interpreted as possible in the situation being modeled.

Fictitious play. Fictitious play involves the choice of strategies at each stage of a game that are optimal against all past choices of the other players.

Final state. A final state is the state induced after all rational moves and countermoves (if any) from the initial state have been made, according to the theory of moves, making it the outcome of the game.

Forward induction. Forward induction is signaling by a player that it is willing and able to cycle indefinitely to assert its moving power.

Game. A game is the totality of rules which describe it. See also Cooperative game; Noncooperative game.

Game configuration. A game configuration is a payoff matrix in which the initial state is not specified.

Game of partial conflict. A game of partial conflict is a variable-sum game in which the players' preferences are not diametrically opposed.

Game theory. Game theory is a mathematical theory of rational strategy selection used to analyze optimal choices in interdependent decision situations; the outcome depends on the choices of two or more players, and each player has preferences for the possible outcomes.

Game of total conflict. A game of total conflict is a constant-sum game in which the best outcome for one player is worst for the other, the next-best outcome for one player is next-worst for the other, and so on.

Game tree. A game tree is a symbolic tree, based on the rules of play of a game, in which the vertices or nodes of the tree represent choice points, and the branches represent alternative courses of action that can be selected by the players.

Generic game. A generic game, by not giving a complete ordering of states but only a partial ordering, subsumes more than one specific ordinal game.

Genuine conflict. A genuine conflict is one in which the three players in the negotiation model all have different most-preferred alternatives.

Impediment. An impediment in a cyclic game occurs when the player with the next move does immediately worse by moving in the direction of the cycling.

Indeterminate state. A state is indeterminate if the outcome induced from it depends on which player moves first (in which case order power is effective).

Information. See Complete information; Perfect information.

Initial state. An initial state is the state in a payoff matrix where play commences.

Intransigence. A player's intransigence is greater the higher it ranks impasse in its preference order (i.e., the more acceptable it finds impasse to be in the negotiation model).

Irrelevant power. Moving power is irrelevant when the outcome induced by one player in a game is better for both players than the outcome that the other player can induce; threat power is irrelevant when either player's possession of it leads to the same outcome.

Lexicographic decision rule. A lexicographic decision rule enables a player to rank states on the basis of a most important criterion ("primary goal"), then a next most important criterion ("secondary goal"), and so on.

Magnanimity. A player is magnanimous when it moves from its best state to its next-best state.

Minimax Theorem. In a two-person constant-sum game, the Minimax Theorem guarantees that each player can ensure at least a certain expected value, called the value of the game, that does not depend on the strategy choice of the other player.

Mixed strategy. A mixed strategy is a strategy that involves a random selection from two or more pure strategies, according to a particular probability distribution.

Move. In a normal-form game, a move is a player's switch from one strategy to another in the payoff matrix.

Moving power. In a cyclic game, moving power is the ability to continue moving when the other player must eventually stop; the player who possesses it uses to induce a preferred outcome.

Nash equilibrium. A Nash equilibrium is a state – or, more properly, the strategies associated with a state – from which no player would have an incentive to depart unilaterally because its departure would immediately lead to a worse, or at least not a better, state.

Noncooperative game. A noncooperative game is a game in which rational players cannot make binding or enforceable agreements but can choose strategies and can move to and from states.

Nonmyopic equilibrium (NME). In a two-person game, a nonmyopic equilibrium is a state from which neither player, anticipating all possible rational moves and countermoves from the initial state, would have an incentive to depart unilaterally because the departure would eventually lead to a worse, or at least not a better, outcome.

Normal (strategic) form. A game is represented in normal (strategic) form when it is described by a payoff matrix in which the players independently choose their strategies. The possible states of the game correspond to the cells of the matrix.

Omniscience. In a two-person game, an omniscient payer is one who can predict the other player's strategy before it is made.

Order power. In a two-person game, order power is the ability of a player to dictate the order of moves in which the players depart from an indeterminate initial state in order to ensure a preferred outcome for itself.

Ordinal game. An ordinal game is a game in which each player can order or rank the states but not necessarily assign numerical payoffs or cardinal utilities to them; when there are no ties in the ranking, the ordering is strict.

Outcome. An outcome is the final state of a game, from which no player chooses to move and at which the players receive their payoffs.

Paradox of omniscience. In a two-person game, a paradox of omniscience occurs when it is in a player's interest to be nonomniscient rather than omniscient.

Paradox of voting. A paradox of voting occurs when no alternative can defeat all the other alternatives in a series of pairwise contests.

Pareto-inferior state. A state is Pareto-inferior if there exists another state that is better for all players, or better for at least one player and not worse for any other players.

Pareto-optimal/superior state. A Pareto-optimal state is one that is not Pareto-inferior. A state is Pareto-superior to a (Pareto-inferior) state if it is better for all players, or better for at least one player and not worse for any other players.

Payoff. A payoff is a measure of the value that a player attaches to a state in a game; usually payoffs are taken to be cardinal utilities, but here they are assumed to be strictly ordinal (ranks from best to worst).

Payoff equivalence. A state is payoff equivalent to an earlier state selected if it gives the player who moves next the same payoff as it had earlier and the same subsequent choices.

Payoff matrix. A payoff matrix is a rectangular array, or matrix, whose entries indicate the payoffs to each player resulting from each of their possible strategy choices.

Payoff termination. Payoff termination occurs when one player receives its best payoff in the move–countermove process and, if it has the next move, terminates play.

Perfect information. A game is one of perfect information when the players know, in the course of play, all the previous choices of the players. See also Complete information.

Platform. A platform is the position of a player (for or against) on two issues in the negotiation model.

Player. See Rational player.

Plurality voting. Plurality voting is a voting procedure in which voters can vote for only one alternative; the alternative with the most votes wins.

Potent threats. Threats are two-player potent when both players possess a threat strategy; one-player potent when one player does; and impotent when no player does.

Power. See Effective power; Irrelevant power; Moving power; Order power; Threat power.

Power ratio. The power ratio is the ratio of the number of wins of a large player to the number of wins of a small player in the negotiation model.

Preference. Preference is a player's ranking of states from best to worst.

Prior communication (PC). Prior communication occurs when a player with threat power, in making a compellent or deterrent threat, communicates its willingness and ability to stay, if necessary, at a Pareto-inferior state.

Pure strategy. A pure strategy is a single specific strategy. See also Mixed strategy.

Rational choice. A rational choice is a choice that leads to a preferred outcome, based on a player's goals.

Rational outcome. See Nash equilibrium; Nonmyopic equilibrium.

Rational player. A rational player is an actor with free will who makes rational choices, in light of the presumed rational choices of other players in a game.

Rational termination. Rational termination is a constraint, assumed in the definition of a nonmyopic equilibrium, that prohibits a player from moving from an initial state unless it leads to a better outcome before cycling.

Rationality rules. Rationality rules specify when players will stay at, or move from, states, taking into account the rational moves of the other players.

Repeated game. A repeated game is a game that comprises repeated play of some component game. See also Supergame.

Repetition termination. Repetition termination occurs when a player reaches a state that is payoff equivalent to a state selected earlier.

Revelation assumption (RA). The revelation assumption in the negotiation model is that players have no information about the preferences of the other players, except as they are revealed in the course of negotiations.

Rules of play. The rules of play of a game describe the possible choices of the players at each stage of play.

Saddlepoint. In a two-person game of total conflict, a saddlepoint is an entry in a payoff matrix at which the row player receives the minimum payoff in its row and the maximum payoff in its column.

Security level. The security level of a player is the best payoff that it can ensure for itself, whatever contingency arises.

Social hierarchy. A social hierarchy is a hierarchy of platforms in the negotiation model such that each platform in the hierarchy can defeat all platforms below it.

Sophisticated voting. Sophisticated voting is voting that involves the successive elimination of dominated strategies, given that other voters act likewise, in a game of complete information.

Stackelberg equilibrium. A Stackelberg equilibrium is a state at which one player (the leader), anticipating the best response of the other player (the follower) to its (the leader's) choice of a strategy, cannot obtain a better outcome for itself by choosing a different strategy.

State. A state is an entry in a payoff matrix from which the players may move. Play of a game starts at an initial state and terminates at a final state, or outcome.

Stoppage. Stoppage occurs when blockage occurs for the first time from some initial state.

Strategy. A strategy is a complete plan that specifies the exact course of action a player will follow, whatever contingency arises.

Supergame. A supergame is a game that comprises infinite repeated play of some component game.

Survivor. A survivor is the state that is selected at any stage as the result of backward induction.

Symmetric game. A symmetric game is one that is strategically the same for all players. In a two-person, normal-form game with ordinal payoffs, a symmetric game is one in which there is an arrangement of the payoffs so that they are the same along the main diagonal, whereas the off-diagonal payoffs are mirror images of each other.

Terminal state. A state is terminal if its choice terminates play of a game.

Theory of moves (TOM). The theory of moves describes optimal strategic calculations in normal-form games in which the players can move and countermove from an initial state.

Threat power. In a two-person game that is repeated, threat power is the ability of a player to threaten a mutually disadvantageous outcome in the single play of a game to deter untoward actions in the future play of this or other games. See also Compellent threat; Deterrent threat.

Threat state/strategy. A threat state is the Pareto-superior state that a threatener promises to implement, by choosing its threat strategy, if the threatened party also agrees to its choice. See also Breakdown state/strategy.

Transitivity. Preferences are transitive if, whenever *x* is preferred to *y* and *y* is preferred to *z*, *x* is preferred to *z*; if this is not the case, preferences are intransitive, leading to cyclical majorities and a paradox of voting.

Trap. A trap is a game in which the players' apparently rational strategies result in a Pareto-inferior outcome, whereby they are all worse off than had they chosen other strategies.

Truel. A truel is the analogue of a duel, in which each of three players can fire or not fire its gun at either of the other two players.

Two-sidedness convention (TSC). The two-sidedness convention describes the conditions under which one player will be magnanimous by moving from a state, even though this move leads to an outcome with a worse payoff for that player. See also Rationality rules; Rules of play.

Two-sidedness rule. The two-sidedness rule describes how players determine whether or not to move from a state on the basis of the other player's rational choices as well as their own.

Undominated strategy. An undominated strategy is a strategy that is neither a dominant nor a dominated strategy.

Utility. Utility is the numerical value, indicating degree of preference, that a player attaches to an outcome.

Value. In two-person constant-sum games, the value is the amount that the players can ensure for themselves by choosing their optimal strategies (which may be mixed) according to classical game theory.

Variable-sum game. A variable-sum game is a partial-conflict game in which the sum of the payoffs to the players in different states is not constant but variable, so the players may gain or lose simultaneously in different states.

Veto. A veto gives a player the ability to block action by a coalition of all the other players.

Weight configuration. A weight configuration is one of the four different sets of weights assigned to the three players in the negotiation model.

Zero-sum game. See Constant-sum (zero-sum) game.

Bibliography

Aaftink, Jan (1989). "Far-Sighted Equilibria in 2×2 Non-Cooperative, Repeated Games." *Theory and Decision* 27, no. 3 (November): 175–92.

Abel, Elie (1966). *The Missile Crisis*. Philadelphia: Lippincott.

Allan, Pierre (1982). *Crisis Bargaining and the Arms Race: A Theoretical Model*. Cambridge, MA: Ballinger.

Allison, Graham T. (1971). *Essence of Decision: Explaining the Cuban Missile Crisis*. Boston: Little, Brown.

Amter, Joseph (1979). *Vietnam Verdict: A Citizen's History*. New York: Continuum.

Ankeny, Nesmith C. (1981). *Poker Strategy: Winning with Game Theory*. New York: Basic.

ApSimon, Hugh (1990). *More Mathematical Byways*. Oxford, UK: Oxford University Press.

Aron, Raymond (1966). *Peace and War*. New York: Doubleday.

Ascherson, Neal (1982). *The Polish Crisis: The Self-Limiting Revolution*. New York: Viking.

Aumann, Robert J. (1989). *Lectures on Game Theory*. Boulder, CO: Westview.

Aumann, Robert J., and Mordecai Kurz (1977). "Power and Taxes." *Econometrica* 45, no. 5 (July): 522–39.

Avenhaus, Rudolf, Steven J. Brams, John Fichtner, and D. Marc Kilgour (1989). "The Probability of Nuclear War." *Journal of Peace Research* 26 (February): 91–9.

Axelrod, Robert (1984). *The Evolution of Cooperation*. New York: Basic.

Baldwin, David A. (1989). *Paradoxes of Power*. New York: Basil Blackwell.

Ball, George (1993). "The Rationalist in Power." *New York Review of Books* 40, no. 8 (April 22): 30–6.

Basar, Tamer, and Geert Jan Olsder (1982). *Dynamic Noncooperative Game Theory*. New York: Academic.

Beck, Nathaniel (1991). "The Illusion of Cycles in International Relations." *International Studies Quarterly* 35, no. 4 (December): 455–76.

Bennett, P. G. (1987). "Beyond Game Theory – Where?" In P. G. Bennett (ed.), *Analyzing Conflict and Its Resolution: Some Mathematical Contributions*. Oxford, UK: Clarendon, pp. 43–69.

Berry, Brian J. L. (1991). *Long-Wave Rhythms in Economic Development and Political Behavior*. Baltimore, MD: Johns Hopkins University Press.

Bialer, Severyn (1981). "Poland and the Soviet Imperium." *Foreign Affairs* 59: 522–39.

Bierman, Scott, and Luis Fernandez (1993). *Game Theory with Economic Applications*. Reading, MA: Addison-Wesley.

Binmore, Ken (1990). *Essays on the Foundations of Game Theory*. Cambridge, MA: Basil Blackwell.

Binmore, Ken (1992). *Fun and Games: A Text on Game Theory*. Lexington, MA: D. C. Heath.

Blight, James G., and David A. Welch (1989). *On the Brink: Americans and Soviets Reexamine the Cuban Missile Crisis*. New York: Hill and Wang.

Bohlen, Celestine (1992). "Yeltsin's Tragic Flaw." *New York Times*, December 13, p. 22.

Bradsher, Keith (1991). "Chip Pact Set by U.S. and Japan." *New York Times*, June 4, pp. D1, D7.

Brams, Steven J. (1975). *Game Theory and Politics*. New York: Free Press.

Brams, Steven J. (1977). "Deception in 2 × 2 Games." *Journal of Peace Science* 2 (Spring): 171–203.

Brams, Steven J. (1978). *The Presidential Election Game*. New Haven, CT: Yale University Press.

Brams, Steven J. (1980). *Biblical Games: A Strategic Analysis of Stories in the Old Testament*. Cambridge, MA: MIT Press.

Brams, Steven J. (1981). "Mathematics and Theology: Game-Theoretic Implications of God's Omniscience." *Mathematics Magazine* 53 (November): 277–82.

Brams, Steven J. (1982a). "Omniscience and Omnipotence: How They May Help – or Hurt – in a Game." *Inquiry* 25, no. 2 (June): 217–31.

Brams, Steven J. (1982b). "A Resolution of the Paradox of Omniscience." In Michael Bradie and Kenneth Sayre (eds.), *Reason and Decision*, Bowling Green Studies in Applied Philsophy, vol. III. Bowling Green, OH: Department of Philosophy, Bowling Green State University, pp. 17–30.

Brams, Steven J. (1983). *Superior Beings: If They Exist, How Would We Know? Game-Theoretic Implications of Omniscience, Omnipotence, Immortality, and Incomprehensibility*. New York: Springer-Verlag.

Brams, Steven J. (1985a). *Rational Politics: Decisions, Games, and Strategy*. Washington, DC: CQ Press. Reprinted by Academic Press, 1989.

Brams, Steven J. (1985b). *Superpower Games: Applying Game Theory to Superpower Conflict*. New Haven, CT: Yale University Press.

Brams, Steven J. (1990). *Negotiation Games: Applying Game Theory to Bargaining and Arbitration*. New York: Routledge.

Brams, Steven J. (1991). "Comment on Fritz W. Scharpf, 'Games Real Actors Could Play.'" *Rationality and Society* 3, no. 2 (April): 252–7.

Brams, Steven J. (1992). "A Generic Negotiation Game." *Journal of Theoretical Politics* 4, no. 1 (January): 53–66.

Brams, Steven J. (1993a). "Game Theory and Literature." *Games and Economic Behavior* (forthcoming).

Brams, Steven J. (1993b). "TOM and the Truel." *Cooperation – or Conflict* 7, no. 1 (January): 2–3

Brams, Steven J. (forthcoming). "Cycles of Conflict." Preprint, Department of Politics, New York University.

Brams, Steven J., and Ann E. Doherty (1993). "Intransigence in Negotiations: The Dynamics of Disagreement." *Journal of Conflict Resolution* (forthcoming).

Brams, Steven J., Ann E. Doherty, and Matthew L. Weidner (1994). "Game Theory and Multilateral Negotiations: The Single European Act and the Uruguay Round." In I. William Zartman (ed.), *Many are Called but Few Choose: The Analysis of Multilateral Negotiations*. San Francisco: Jossey-Bass (forthcoming).

Brams, Steven J., Dan S. Felsenthal, and Zeev Maoz (1986). "New Chairman Paradoxes." In Andreas Diekmann and Peter Mitter (eds.), *Paradoxical Effects of Social Behavior: Essays in Honor of Anatol Rapoport*. Heidelberg, Germany: Physica-Verlag, pp. 243–56.

Brams, Steven J., Dan S. Felsenthal, and Zeev Maoz (1988). "Chairman Paradoxes under Approval Voting." In Gerald Eberlein and Hal Berghel (eds.), *Theory and Decision: Essays in Honor of Werner Leinfellner*. Dordrecht, Holland: D. Reidel.

Brams, Steven J., and Peter C. Fishburn (1983). *Approval Voting*. Cambridge, MA: Birkhäuser Boston.

Brams, Steven J., and Marek P. Hessel (1982). "Absorbing Outcomes in 2×2 Games." *Behavioral Science* 27, no. 4 (October): 393–401.

Brams, Steven J., and Marek P. Hessel (1983). "Staying Power in 2×2 Games." *Theory and Decision* 15, no. 3 (September): 279–302.

Brams, Steven J., and Marek P. Hessel (1984). "Threat Power in Sequential Games." *International Studies Quarterly* 28, no. 1 (March): 15–36.

Brams, Steven J., and D. Marc Kilgour (1988). *Game Theory and National Security*. New York: Basil Blackwell.

Brams, Steven J., and Walter Mattli (1993)."Theory of Moves: Overview and Examples." *Conflict Management and Peace Science* 12, no. 2: 1–39.

Brams, Steven J., and Ben D. Mor (1993). "When Is It Rational to Be Magnanimous in Victory?" *Rationality and Society* (forthcoming).

Brams, Steven J., and Donald Wittman (1981). "Nonmyopic Equilibria in 2×2 Games." *Conflict Management and Peace Science* 6, no. 1 (Fall): 39–62.

Brams, Steven J., and Frank C. Zagare (1977). "Deception in Simple Voting Games." *Social Science Research* 6 (September): 257–72.

Brams, Steven J., and Frank C. Zagare (1981). "Double Deception: Two against One in Three-Person Games." *Theory and Decision* 13 (March): 81–90.

Brugoni, Dina A. (1992). *Eyeball to Eyeball: The Inside Story of the Cuban Missile Crisis*. New York: Random House.

Brune, Lester H. (1985). *The Missile Crisis of October 1962: A Review of Issues and References*. Claremont, CA: Regina.

Buber, Martin (1958). *I and Thou*, trans. by Ronald Gregor Smith, 2nd edn. New York: Scribner's.

Carlsson, Hans, and Eric van Damme (1991). "Equilibrium Selection in Stag Hunt Games." Preprint, Center for Economic Research, Tilburg University, The Netherlands.

Carter, Bill (1992). "Is It Jay? David? No Matter; Winner May be Ted Koppel." *New York Times*, December 28, p. D6.

Carter, Jimmy (1982). *Keeping Faith: Memoirs of a President*. New York: Bantam.

Chayes, Abram (1974). *The Cuban Missile Crisis: International Crises and the Role of Law*. New York: Oxford University Press.

Cioffi-Revilla, Claudio (1983). "A Probability Model of Credibility." *Journal of Conflict Resolution* 27, no. 1 (March): 73–108.

Clausewitz, Karl von (1832). *On War*, edited by Anatol Rapoport. New York: Penguin (1966).

Clines, Francis X. (1991). "Gorbachev Recounts Telling Plotters: To Hell with You." *New York Times*, August 31, p. A13.

Cohen, Raymond (1991). *Negotiating Across Cultures*. Washington, DC: U.S. Institute of Peace.

Coll, Juan Carlos Martinez, and Jack Hirshleifer (1991). "The Limits of Reciprocity: Solution Concepts and Reactive Strategies in Evolutionary Equilibrium Models." *Rationality and Society* 3, no. 1 (January): 35–64.

Conklin, John E. (1972). *Robbery and the Criminal Justice System*. Philadelphia: Lippincott.

Cross, John G., and Melvin J. Guyer (1980). *Social Traps*. Ann Arbor, MI: University of Michigan Press.

Dalkey, Norman C. (1981). "A Case Study of a Decision Analysis: Hamlet's Soliloquy." *Interfaces* 11, no. 5 (October): 45–9.

Daniel, Donald C., and Katherine L. Herbig (eds.) (1982). *Strategic Military Deception*. New York: Pergamon.

Davis, Morton D. (1970). *Game Theory: A Nontechnical Introduction*. New York: Basic (2nd edn, 1983).

Davis, Morton D. (1990). Private communication (September 6).

De, Mitali, Keith W. Hipel, and D. Marc Kilgour (1990). "Algorithms for Hierarchical Power." *Applied Mathematics and Computation* 39: 21–36.

Detzer, David. (1979). *The Brink: Story of the Cuban Missile Crisis*. New York: Crowell.

Dimand, Robert W., and Mary Ann Dimand (1992). "The Early History of the Theory of Strategic Games from Waldegrave to Borel." In E. Roy Weintraub (ed.), *Toward a History of Game Theory*. Durham, NC: Duke University Press, pp. 15–27.

Dinerstein, Herbert (1976). *The Making of the Cuban Missile Crisis, October 1962*. Baltimore: Johns Hopkins University Press.

Divine, Robert A. (ed.) (1971). *The Cuban Missile Crisis*. Chicago: Quadrangle.

Dixit, Avinash, and Barry Nalebuff (1991). *Thinking Strategically: The Competitive Edge in Business, Politics, and Everyday Life*. New York: W. W. Norton.

Eban, Abba (1977). *An Autobiography*. New York: Random House.

Epstein, Richard A. (1967). *The Theory of Gambling and Statistical Logic*. New York: Academic.

Farnsworth, Clyde H. (1990). "Trade Talks Will Continue in Informal Settings." *New York Times*, December 14, p. D2.

Farquharson, Robin (1971). *Theory of Voting*. New Haven, CT: Yale University Press.

Faulkner, William (1950). *Light in August*. New York: Random House.

Fishburn, Peter C. (1974). "Lexicographic Orders, Utilities and Decision Rules: A Survey." *Management Science* 20, no. 11 (July): 1442–71.

Fisher, Franklin M. (1989). "Games Economists Play: A Noncooperative View." *RAND Journal of Economics* 20, no. 1 (Spring 1989): 113–24.

Fisher, Roger, and William Ury (1983). *Getting to Yes*. New York: Penguin.

Fowler, James W. (1981). *Stages of Faith: The Psychology of Human Development and the Quest for Meaning*. San Francisco: Harper and Row.

Frank, Robert H. (1988). *Passions within Reason: The Strategic Role of the Emotions*. New York: W. W. Norton.

Fraser, Niall M., and Keith W. Hipel (1982–3). "Dynamic Modeling of the Cuban Missile Crisis." *Conflict Management and Peace Science* 6, no. 2 (Spring): 1–18.

Friedman, James W. (1990). *Game Theory with Applications to Economics*, 2nd edn. New York: Oxford University Press.

Fudenberg, Drew, and Jean Tirole (1991). *Game Theory*. Cambridge, MA: MIT Press.

Gardner, Roy, and Elinor Ostrom (1991). "Rules and Games." *Public Choice* 70, no. 2 (May): 121–49.

Garthoff, Raymond L. (1989). *Reflections on the Cuban Missile Crisis*, rev. edn. Washington, DC: Brookings.

Geanakopolos, John, David Pearce, and Ennio Stacchetti (1989). "Psychological Games and Sequential Rationality." *Games and Economic Beahvior* 1, no. 1 (March): 60–79.

Gibbons, Robert (1992). *Game Theory for Applied Economists*. Princeton, NJ: Princeton University Press.

Gilboa, Itzhak, and David Schmeidler (1988). "Information Dependent Games: Can Common Sense Be Common Knowledge?" *Economic Letters* 27: 215–21.

Goldstein, Joshua S. (1988). *Long Cycles: Prosperity and War in the Modern Age*. New Haven, CT: Yale University Press.

Goldstein, Joshua S. (1991). "The Possibility of Cycles in International Relations." *International Studies Quarterly* 35, no. 4 (December): 477–80.

Grandmont, Jean-Michel (ed.) (1988). *Temporary Equilibrium: Selected Readings*. San Diego, CA: Academic.

Greenberg, Joseph (1990). *The Theory of Social Situations: An Alternative Game-Theoretic Approach*. Cambridge, UK: Cambridge University Press.

Güth, Werner (1991). "Game Theory's Basic Question: Who Is a Player?" *Journal of Theoretical Politics* 3, no. 4 (October): 403–35.

Güth, Werner, and Brigitte Kalkofen (1989). *Unique Solution to Strategic Games*. Berlin: Springer-Verlag.

Hamilton, Jonathan H., and Steven M. Slutsky (1993). "Endogenizing the Order of Moves in Matrix Games." *Theory and Decision* 34, no. 1 (January): 47–62.

Hanson, Norwood Russell (1971). *What I Don't Believe and Other Essays*, ed. by Stephen Toulmin and Harry Woolf. Dordrecht, Holland: D. Reidel.

Harkabi, Yehoshafat (1972). *Arab Attitudes toward Israel*, trans. by Misha Louvish. New York: Hart.

Harsanyi, John C. (1977). *Rational Behavior and Bargaining Equilibrium in Games and Social Situations*. New York: Cambridge University Press.

Harsanyi, John C., and Reinhard Selten (1988). *A General Theory of Equilibrium Selection in Games*. Cambridge, MA: MIT Press.

Haywood, O. G., Jr. (1954). "Military Decision and Game Theory." *Operations Research* 2 (November): 365–85.

Headley, J. T. (1863). *The Great Rebellion: A History of the Civil War in the United States*, vol. I. Hartford, CT: Hulburt, Williams & Company.

"Hedging" (1990). *Economist*, October 13, p. 74.

Heims, Steve J. (1980). *John von Neumann and Norbert Weiner: From Mathematics to the Technologies of Life and Death*. Cambridge, MA: MIT Press.

Herring, George C. (1979). *America's Longest War: The United States and Vietnam, 1950–1975*. New York: Wiley.

Higgott, Richard A., and Andrew Fenton Cooper (1989). "Middle Power Leadership and Coalition Building: Australia, the Cairns Group, and the Uruguay Round of Trade Negotiations." *International Organization* 44, no. 4 (Autumn): 589–632.

Hillinger, Claude (1971). "Voting on Issues and on Platforms." *Behavioral Science* 16, no. 6 (November): 564–6.

Hipel, Keith W., Muhong Wang, and Niall M. Fraser (1988). "Hypergame Analysis of the Falkland/Malvinas Conflict." *International Studies Quarterly* 32, no. 3 (September): 335–58.

Hirshleifer, Jack (1985). "Protocol, Payoff, and Equilibrium: Game Theory and Social Modeling." Preprint, Department of Economics, University of California, Los Angeles.

Ho, Y. C., and B. Tolwinski (1982). "Credibility and Rationality of Players' Strategies in Multilevel Games." *Proceedings of the 21st Conference on Decision and Control* (December 8–10).

Holsti, Ole R., Richard A. Brody, and Robert C. North (1964). "Measuring Affect and Action in International Reaction Models: Empirical Materials from the 1962 Cuban Missile Crisis." *Journal of Peace Research* 1: 170–89.

Howard, Nigel (1971). *Paradoxes of Rationality: Theory of Metagames and Political Behavior*. Cambridge, MA: MIT Press.

Howard, Nigel (1992). "Dramatic Transformations of the Core." Preprint, Nigel Howard Systems, Birmingham, UK.

Howard, Nigel (1993). "Response: The Truel Dramatised." *Cooperation – or Conflict* 7, no. 1 (January): 3–4

Howard, Nigel, Peter Bennett, Jim Bryant, and Morris Bradley (1993). "Manifesto for a Theory of Drama and Irrational Choice." *Systems Practice* 6, no. 4: 429–34.

Isaacs, Arnold R. (1983). *Without Honor: Defeat in Vietnam and Cambodia*. Baltimore, MD: Johns Hopkins University Press.

Kadane, Joseph B. (1972). "On Division of the Question." *Public Choice* 13 (Fall): 47–54.

Karsh, Efraim (1989). "Military Lessons of the Iran–Iraq War." *Orbis* 33, no. 2 (Spring): 209–23.

Kennedy, Robert F. (1969). *Thirteen Days: A Memoir of the Cuban Missile Crisis*. New York: Norton.

Kilgour, D. Marc (1973). "The Simultaneous Truel." *International Journal of Game Theory* 1, no. 4: 229–42.

Kilgour, D. Marc (1975). "The Sequential Truel." *International Journal of Game Theory* 4, no. 3: 151–74.

Kilgour, D. Marc (1978). "Equilibrium Points of Infinite Sequential Truels." *International Journal of Game Theory* 6, no. 3: 167–80.

Kilgour, D. Marc (1984). "Equilibria for Far-sighted Players." *Theory and Decision* 16, no. 2 (March): 135–57.

Kilgour, D. Marc (1985). "Anticipation and Stability in Two-Person Non-Cooperative Games." In Urs Luterbacher and Michael D. Ward (eds.), *Dynamic Models of International Conflict*. Boulder, CO: Lynne Rienner, pp. 26–51.

Kilgour, D. Marc, Mitali De, and Keith W. Hipel (1987). "Conflict Analysis Using Staying Power." *Proceedings of the 1986 IEEE International Conference on Systems, Man and Cybernetics* (Atlanta, GA).

Kilgour, D. Marc, and Frank C. Zagare (1987). "Holding Power in Sequential Games." *International Interactions* 13, no. 2: 91–114.

Kinzer, Stephen (1993). "Honecker Urged Poland Invasion Amid Furor over Solidarity in 80." *New York Times*, January 12, p. A7.

Kissinger, Henry A. (1964). *A World Restored*. New York: Grosset and Dunlap.

Kolakowski, Leszek (1982). *Religion*. New York: Oxford University Press.

Kreps, David M. (1990). *Game Theory and Economic Modelling*. New York: Oxford University Press.

Kuron, Jacek (1981). Interview, *Telos* 47: 93–7.

Langlois, Jean-Pierre P. (1991). "Rational Deterrence and Crisis Stability." *American Journal of Political Science* 35, no. 4 (November): 801–32.

Langlois, Jean-Pierre P. (1992). "Decision Systems Analysis of Crisis Behavior." Preprint, Department of Mathematics, San Francisco State University.

"Last Word on the October Surprise?" (1993). *New York Times*, January 24, p. E17.

Lewis, Neil A. (1992). "Panel Rejects Theory Bush Met Iranians in Paris in 80." *New York Times*, July 2, p. A16.

Lewis, Neil A. (1993). "House Inquiry Finds No Evidence of Deal on Hostages in 1980." *New York Times*, January 13, pp. A1, A19.

Lewy, Guenter (1978). *America in Vietnam*. New York: Oxford University Press.

Luce, R. Duncan, and Howard Raiffa (1957). *Games and Decisions: Introduction and Critical Survey*. New York: Wiley.

McMillan, John (1992). *Games, Strategies, and Managers*. New York: Oxford University Press.

Macrae, Norman (1992). *John von Neumann*. New York: Pantheon.

Mailath, George J., Larry Samuelson, and Jeroen Swinkels (1993). "Extensive Form Reasoning in Normal Form Games." *Econometrica* 61, no. 2 (March): 273–302.

Maoz, Zeev (1983). "Resolve, Capabilities, and the Outcomes of Interstate Disputes." *Journal of Conflict Resolution* 27, no. 2 (June): 195–229.

Maoz, Zeev (1984). "Peace by Empire? Conflict Outcomes and International Stability, 1816–1979." *Journal of Peace Research* 21, no. 3: 227–41.

Maoz, Zeev (1990). *Paradoxes of War: On the Art of National Self-Entrapment*. Boston: Unwin Hyman.

Marschak, Thomas, and Reinhard Selten (1978). "Restabilizing Responses, Inertia Supergames, and Oligopolistic Equilibria." *Quarterly Journal of Economics* 42: 71–93.

Matsui, Akihiko, and Kiminori Matsuyama (1991). "An Approach to Equilibrium Selection." Preprint, Center for Mathematical Studies in Economics and Management Science, Northwestern University.

Mattli, Walter (1992). "The Iranian Hostage Crisis: A Game-Theoretic Sketch." Preprint, Department of Political Science, University of Chicago.

Mertens, Jean-Francois (1991). "Equilibrium and Rationality: Context and History-Dependence." In Kenneth J. Arrow (ed.), *Issues in Contemporary Economics: Markets and Welfare*, vol. I. New York: New York University Press, pp. 198–211.

Modis, Theodore (1992). *Predictions: Society's Telltale Signature Reveals the Past and Forecasts the Future*. New York: Simon and Schuster.

Moldovanu, Benny, and Eyal Winter (n.d.). "Order Independent Equilibria." Preprint, Department of Economics, University of Bonn.

Mor, Ben D. (1993). *Decision and Interaction in Crisis: A Model of International Crisis Behavior*. Westport, CT: Greenwood.

Morgenstern, Oskar (1928). *Wirtschaftsprognose, Eine Untersuchung ihrer Voraussetzungen und Möglichkeiten* (*Economic Prediction: An Examination of Its Conditions and Possibilities*). Vienna: J. Springer.

Morgenstern, Oskar (1935). "Volkommene Voraussicht und Wirtschaftliches Gleichgewicht." *Zeitschrift für Nationalökonomie* 6, part 3 (August): 337–57, trans. as "Perfect Foresight and Economic Equilibrium." In Andrew Schotter (ed.) (1976), *Selected Writings of Oskar Morgenstern*. New York: New York University Press, pp. 169–83.

Moulin, Hervé (1981). "Deterrence and Cooperation: A Classification of Two-Person Games." *European Economic Review* 15: 179–93.

Moulin, Hervé (1986). *Game Theory for the Social Sciences*, 2nd rev. edn. New York: New York University Press.

Mueller, John (1989). *Retreat from Doomsday: The Obsolescence of Major War*. New York: Basic.

Muzzio, Douglas (1982). *Watergate Games: Strategies, Choices, Outcomes*. New York: New York University Press.

Myerson, Roger B. (1991). *Game Theory: Analysis of Conflict*. Cambridge, MA: Harvard University Press.

Nash, John (1951). "Non-cooperative Games." *Annals of Mathematics* 54: 286–95.

Orbell, John (n.d.). "The Tragedy of Hamlet, Prince of Denmark: Decision-Making under Risk and Uncertainty." Preprint, Department of Political Science, Universitiy of Oregon.

Ordeshook, Peter C. (1986). *Game Theory and Political Theory*. Cambridge, UK: Cambridge University Press.

Oren, Nissan (1982). "Prudence in Victory." In Nissan Oren (ed.), *Termination of Wars*. Jerusalem: Magnes, pp. 147–63.

Oxley, Alan (1990). *The Challenge of Free Trade*. London: Harvester Wheatsheaf.

Pachter, Henry M. (1963). *Collision Course: The Cuban Missile Crisis and Coexistence*. New York: Praeger.

Palmer, Dave Richard (1978). *Summons of the Trumpet*. San Rafael, CA: Presidio.

Papp, Daniel S. (1981). *Vietnam: The View from Moscow, Peking, Washington*. Jefferson, NC: McFarland.

Passel, Peter (1990). "Adding Up the World Trade Talks: Fail Now, Pay Later." *New York Times*, December 16, p. E3.

Passel, Peter (1991). "Second Thoughts on Mexico Trade." *New York Times*, May 15, p. D2.

Pear, Robert (1992). "The Cuba Missile Crisis: Kennedy Left a Loophole." *New York Times* (January 7), p. A5.

Porter, Gareth (1975). *A Peace Denied: The United States, Vietnam, the Paris Agreement*. Bloomington, IN: Indiana University Press.

Poundstone, William (1992). *Prisoners' Dilemma*. New York: Doubleday.

Powell, Robert (1989). *Nuclear Deterrence Theory: The Search for Credibility*. Cambridge, UK: Cambridge University Press.

Pratt, Michael (1980). *Mugging as a Social Problem*. London: Routledge and Kegan Paul.

Prestowitz, Clyde (1991). "Life after GATT: More Trade Is Better Than Free Trade." *Technology Review* 94, no. 3 (April): 21–7.

The Prophets (1978). Philadelphia: Jewish Publication Society.

Quandt, William B. (1986). *Camp David: Peacemaking and Politics*. Washington, DC: Brookings.

Raiffa, Howard (1982). *The Art and Science of Negotiation*. Cambridge, MA: Harvard University Press.

Rapoport, Anatol, and Melvin Guyer (1966). "A Taxonomy of 2×2 Games." *General Systems: Yearbook of the Society for General Systems Research* 11: 203–14.

Rasmusen, Eric (1989). *Games and Information: An Introduction to Game Theory*. Oxford, UK: Basil Blackwell.

Ravid, Itzhak (1990). "Military Decision, Game Theory and Intelligence: An Anecdote." *Operations Research* 39, no. 2 (March–April): 260–4.

Richerson, Peter J., and Robert Boyd (1987). "Simple Models of Complex Phenomena: The Case of Cultural Evolution." In John Dupré (ed.), *The Latest on the Best: Essays on Evolution and Optimality*. Cambridge, MA: MIT Press, pp. 27–52.

Rosecrance, Richard (1987). "Long Cycle Theory and International Relations." *International Organization* 41: 283–302.

Rubinstein, Ariel (1991). "Comments on the Interpretation of Game Theory." *Econometrica* 59, no. 4 (July): 909–24.

Sachar, Howard M. (1979). *A History of Israel: From the Rise of Zionism to Our Time*. New York: Knopf.

Samuelson, Larry, and Jiambo Zhang (1992). "Evolutionary Stability in Dynamic Games." *Journal of Economic Theory* 57, no. 2 (August): 363–91.

Sanger, David E. (1991). "U.S. Companies in Japan Say Things Aren't So Bad." *New York Times*, June 11, pp. A1, D6.

Saunders, Harold (1985). "The Crisis Begins." In Warren Christopher (ed.), *American Hostages in Iran: The Conduct of a Crisis*. New Haven, CT: Yale University Press, pp. 35–71.

Schelling, Thomas C. (1966). *Arms and Influence*. New Haven, CT: Yale University Press.

Sebenius, James K. (1984). *Negotiating the Law of the Sea*. Cambridge, MA:Harvard University Press.

Selten, Reinhard (1975). "Reexamination of the Perfectness Concept for Equilibrium Points in Extensive Games." *International Journal of Game Theory* 4, no. 1: 25–55.

Sexton, Thomas R., and Dennis R. Young (1985). "Game Tree Analysis of International Crises." *Journal of Policy Analysis and Management* 4, no. 3: 354–69.

Shakespeare, William (1958). *The Tragedy of Hamlet, Prince of Denmark*, The Folger Library Shakespeare. New York: Pocket Books.

Shapley, L. S. (1964). "Some Topics in Two-Person Games." In M. Dresher, L. S. Shapley, and A. W. Tucker (eds.), *Advances in Game Theory, Annals of Mathematics Studies* 52: 1–28.

Sharp, U. S. G. (1978). *Strategy for Defeat: Vietnam in Retrospect*. San Rafael, CA: Presidio.

Shubik, Martin (1982). *Game Theory in the Social Sciences: Concepts and Solutions*. Cambridge, MA: MIT Press.

Shubik, Martin (1984). *A Game-Theoretic Approach to Political Economy*. Cambridge, MA: MIT Press.

Sick, Gary (1985a). *All Fall Down*. New York: Penguin.

Sick, Gary (1985b). "Military Options and Constraints." In Warren Christopher (ed.), *American Hostages in Iran: The Conduct of a Crisis*. New Haven, CT: Yale University Press, pp. 144–72.

Sick, Gary (1991). *October Surprise: America's Hostages in Iran and the Election of Ronald Reagan*. New York: Random House.

Silk, Leonard (1991). "Trade Bloc War? Concern Grows." *New York Times*, April 26, p. D2.

Silverman, David L. (1991). *Your Move: Logic, Math and Word Puzzles for Enthusiasts*. New York: Dover.

Sjöstedt, Gunnar (1990). "Multilateral Negotiations: The Story of the Uruguay Round." Preprint, Swedish Institute of International Affairs (December).

Skyrms, Brian (1990). *The Dynamics of Rational Deliberation*. Cambridge, MA: Harvard University Press.

Snyder, Glenn H., and Paul Diesing (1977). *Conflict among Nations: Bargaining, Decision Making, and Systems Structure in International Crises*. Princeton, NJ: Princeton University Press.

Sorensen, Theodore C. (1965). *Kennedy*. New York: Harper and Row.

Stackelberg, Heinrich von (1934). *Marktform und Gleichgewicht*. Berlin: J. Springer. Trans. by Alan Peacock as *The Theory of the Market Economy*. London: William Hodge, 1952.

Stampp, Kenneth M. (1950). *And the War Came: The North and the Secession Crisis, 1860–61*. Baton Rouge, LA: Louisiana State University Press.

Staudenraus, P. J. (ed.) (1963). *The Secession Crisis, 1860–61*. Chicago: Rand McNally.

Swinburne, Richard (1981). *Faith and Reason*. Oxford: Clarendon.

Szafar, Tadeusz (1981). "Brinkmanship in Poland." *Problems of Communism* 30 (May/June): 75–81.

Thomas, L. C. (1987). "Using Game Theory and Its Extensions to Model Conflict." In P. G. Bennett (ed.), *Analyzing Conflict and Its Resolution: Some Mathematical Contributions*. Oxford, UK: Clarendon, pp. 3–22.

Thompson, James Clay (1980). *Rolling Thunder: Understanding Policy and Program Failure*. Chapel Hill, NC: University of North Carolina Press.

Thompson, Robert Smith (1992). *The Missiles of October: The Declassified Story of John F. Kennedy and the Cuban Missile Crisis*. New York: Simon and Schuster.

Thompson, William R. (1992). "Systemic Leadership and Growth Waves in the Long Run." *International Studies Quarterly* 36, no. 1 (March): 25–48.

Tolchin, Martin (1992). "U.S. Underestimated Soviet Force in Cuba During '62 Missile Crisis." *New York Times*, January 15, p. A11.

The Torah: The Five Books of Moses, 2nd edn (1967). Philadelphia: Jewish Publication Society.

Touval, Saadia (1982). *The Peace Brokers: Mediators in the Arab–Israeli Conflict, 1948–1979*. Princeton, NJ: Princeton University Press.

Uchitelle, Louis (1991). "Blocs Seen Replacing Free Trade." *New York Times*, August 26, pp. D1, D4.

Vance, Cyrus (1983). *Hard Choices: Critical Years in America's Foreign Policy*. New York: Simon and Schuster.

van Damme, Eric (1991a). "Equilibrium Selection in 2×2 Games." *Revista Espanola de Economia* 8, no. 1: 37–52.

van Damme, Eric (1991b). *Stability and Perfection of Nash Equilibria*, 2nd edn. Heidelberg, Germany: Springer-Verlag.

von Neumann, John (1928). "Zur Theorie der Geselleschaftsspiele." *Mathematische Annalen* 100: 295–300, trans. as "On the Theory of Games of Strategy." In A. W. Tucker and R. D. Luce (eds.), *Contributions to the Theory of Games 4, Annals of Mathematics Studies* 40 (1959): 13–42.

von Neumann, John, and Oskar Morgenstern (1953). *Theory of Games and Economic Behavior*, 3rd edn. Princeton, NJ: Princeton University Press.

Watt, Richard M. (1982). "Polish Possibilities." *New York Times Book Review*, April 25, pp. 11, 19.

Weintal, Edward, and Charles Bartlett (1967). *Facing the Brink: An Intimate Study of Crisis Diplomacy*. New York: Scribner.

Weintraub, E. Roy (ed.) (1992). *Toward a History of Game Theory*. Durham, NC: Duke University Press.

Whitaker, Catherine J. (1989). "The Redesigned National Crime Survey: Selected New Data." Special Report, Bureau of Justice Statistics, U.S. Department of Justice (January).

Wilson, Robert (1985). "Reputation in Games and Markets." In Alvin E. Roth (ed.), *Game-Theoretic Models of Bargaining*. New York: Cambridge University Press, pp. 27–62.

Wilson, Robert (1989). "Deterrence in Oligopolistic Competition." In James L. Stern, Robert Axelrod, Robert Jervis, and Roy Radner (eds.), *Perspectives on Deterrence*. New York: Oxford University Press, pp. 157–90.

Winham, Gilbert R., and Karin L. Kizer (1990). "The Uruguay Round Mid-Term Review: A Case Study of Multilateral Trade Negotiations." Preprint (March 14).

Yaniv, Avner (1987). *Deterrence without the Bomb*. Lexington, MA: Lexington.

Yarbrough, Beth V., and Robert M. (1992). *Cooperation and Governance in International Trade: The Strategic Organizational Approach*. Princeton, NJ: Princeton University Press.

Young, Peyton (ed.) (1991). *Negotiation Analysis*. Ann Arbor, MI: University of Michigan Press.

Zagare, Frank C. (1977). "A Game-Theoretic Analysis of the Vietnam Negotiations: Preferences and Strategies 1968–73." *Journal of Conflict Resolution* 21, no. 4 (December): 663–84.

Zagare, Frank C. (1979). "The Geneva Conference of 1954: A Case of Tacit Deception." *International Studies Quarterly* 23, no. 3 (September): 390–411.

Zagare, Frank C. (1981). "Nonmyopic Equilibria and the Middle East Crisis of 1967." *Conflict Management and Peace Science* 5 (Spring): 139–62.
Zagare, Frank C. (1983). "A Game-Theoretic Evaluation of the Cease-Fire Alert Decision of 1973." *Journal of Peace Research* 20, no. 1: 73–86.
Zagare, Frank C. (1984). "Limited-Move Equilibria in 2 × 2 Games." *Theory and Decision* 16, no. 1 (January): 1–19.
Zagare, Frank C. (1987). *The Dynamics of Deterrence*. Chicago: University of Chicago Press.
Ziegler, David W. (1987). *War, Peace, and International Politics*, 4th edn. Glenview, IL: Scott, Foresman.

Index